'Till The Peoples All Are One'

Darwin's Unitarian Connections

Clifford M. Reed

The Lindsey Press
London

Published by the Lindsey Press
on behalf of the General Assembly of Unitarian
and Free Christian Churches
Essex Hall, 1–6 Essex Street, London WC2R 3HY, UK

ISBN 978-0-85319-082-0

Designed and typeset by Garth Stewart, Oxford

Printed and bound in the United Kingdom by
Lightning Source, Milton Keynes

Contents

About the author

Clifford Reed is a lifelong Unitarian and has been the minister at the Unitarian Meeting House in Ipswich since 1976. A past President of the Unitarian and Free Christian Churches, he has served the denomination in many capacities at local, district, national, and international levels. He has written and edited books on Unitarian thought, among them the popular introduction, *Unitarian? What's That?* His published work also includes devotional material, such as *Spirit of Time and Place* and *Sacred Earth*. He has contributed a number of hymns to recent Unitarian hymnals.

This little book began life as a lecture given at the Ipswich Meeting House as part of the Darwin bicentennial celebrations in 2009. Clifford Reed has had a life-long active interest in the natural world, an interest that is both spiritual ('in the best rational, Unitarian sense!') and Darwinian.

Preface

Even though it left many questions unanswered at the time of its publication, few people today would doubt the basic truth of Darwin's theory of evolution by means of natural selection – few, that is, apart from those who, for ideological reasons, are determined to disbelieve it. This opposition to Darwin comes, in the main, from people who are known as 'religious fundamentalists'. And because they can be very vocal and, in some places, very influential, the idea that religion is in conflict with science in general and with Darwin in particular has, if anything, gained currency in recent years. But in fact most Christian churches were already coming to terms with Darwin by the time he died and were more concerned with adapting their doctrines accordingly than with fighting a rearguard action against the theory of evolution. But no denomination was more at ease with Darwin than the Unitarians: 'There is no such Unitarian as science', declared one nineteenth-century Unitarian minister. 'There is no better Unitarian literature than Darwin's "Origin of Species".'

Darwin himself was no more dogmatic on the subject of religion than he was on matters of science – except where he regarded its teachings as cruel, inhumane, and manifestly untrue. Just as he never presented his science as dogma and was the first to point out the gaps and weak points in his own arguments, so in religion he moved steadily to a position where, on matters theological, he was content not to know the answers to the supposedly 'big' questions. This did not mean that he ceased to be religious, however, as is so often claimed. Rather his religion was concerned with the 'small' questions, such as how we can remove the fetters of ignorance and superstition, how we can liberate the human mind and spirit, and how we can release the human potential to build a better world. In such a religion there could be no conflict with science, because both exist to serve the cause of human progress.

Although he was baptised in one Anglican church and buried in another, Darwin's religious journey began in Unitarianism and, arguably, found its fulfilment there too. And, in all the years between his baptism and funeral, there were few times, if any, when he ceased to be connected with Unitarians of one sort or another, or to be influenced by them. The connections that will be considered in the following chapters were not solely his innumerable ties of kinship and friendship within the complex web of nineteenth-century Unitarianism, but also the connections between Darwin and the causes that he shared with Unitarians – such as the long struggle against slavery; the scientific quest for knowledge and understanding; the commitment to human advancement; and the attempt to make sense of our sometimes painful existence in a capricious universe. Unitarians were not, of course, unique in being concerned with these things, but they could approach them without the straitjackets of either dogmatic religion or dogmatic secularism. Reason and faith, science and religion, were not – and are not – seen as opposites, but rather as being in ceaseless and creative dialogue. It is with Darwin's life-long relationship with this peculiar brand of liberal, non-dogmatic, and open-ended faith community that the following chapters are concerned.

Clifford Reed

January 2011

Note: All page references in the ensuing text relate to the editions listed under 'Sources'.

1 Ambivalent beginnings: Deists, 'Lunaticks', and Unitarians

Charles Robert Darwin was born in Shrewsbury on 12 February 1809, and from the very beginning of his life-story it is possible to detect some of the complications that bedevil any examination of his relationship with Unitarianism.

Charles's father, **Dr Robert Darwin (1766–1848)**, is usually characterised as a 'freethinker', like his paternal grandfather, **Dr Erasmus Darwin (1731–1802)**, an advocate of rational and scientific enquiry, free from the restricting influence of the Church. But Erasmus should also properly be called a 'Deist'. Although decidedly sceptical about Christian doctrine and the authority of the Bible, Erasmus Darwin did believe in a Creator God, in the sense of 'a First Cause, a Being of Beings'[1] which he described as 'Potent-power, all-great, all-good'.[2] This remote Deity does not intervene directly in the universe that he has set in motion, unlike the God of Christianity (including Christianity in its contemporary eighteenth-century Unitarian form), but he has ordered the universe benignly. As Erasmus Darwin wrote in a letter:[3]

> That there exists a superior ENS ENTIUM,[4] which formed these wonderful creatures, is mathematical demonstration. That HE influences things by a particular providence, is not so evident. The probability, according to my notion, is against it.

---

1 Jenny Uglow, *The Lunar Men*, p. 39.
2 Desmond and Moore, *Darwin*, p. 6.
3 Uglow, op. cit., p. 39.
4 *Ens entium*: literally 'Being of Beings'.

Portrait of Erasmus Darwin by Joseph Wright of Derby (1792)

In the Preface to Volume One of his major work, *Zoonomia*, Erasmus Darwin gives another clear statement of this belief and adds a further dimension to it.

*The great CREATOR of all things has infinitely diversified the works of
his hands, but has at the same time stamped a certain similitude on the
features of nature, that demonstrates to us, that the whole is one family
of one parent.*[5]

This surely formed an influential aspect of the intellectual environ-
ment within which his grandson was to grow up. Later in the book,
Erasmus Darwin makes this almost lyrical statement concerning the
Deist position:

*This perpetual chain of causes and effects, whose first link is riveted to
the throne of God, divides itself into innumerable diverging branches,
which ... permeate the most minute and most remote extremities of the
system, diffusing motion and sensation to the whole. As every cause is
superior in power to the effect, which it has produced, so our idea of the
Almighty Creator becomes more elevated and sublime, as we trace the
operations of nature from cause to cause, climbing up the links of these
chains of being, till we ascend to the Great Source of all things.*[6]

It was Erasmus Darwin's view that, far from conflicting with belief in
God, '... the modern discoveries in chemistry and in geology ... as well as
those in astronomy, which dignify the present age, contribute to enlarge
and amplify our ideas of the power of the Great First Cause'.[7]

This was, of course, the Deist version of God, rather than an 'orthodox' Christian one, but nevertheless Erasmus Darwin is at pains to
distance himself from the atheist position that he is sometimes accused
of holding. Had the 'ancient philosophers' been prepared to attribute
the combination of atoms that formed the world to 'immutable proper-
ties received from the hand of the Creator', rather than to 'blind chance',
then, Darwin writes, 'the doctrine of atoms ... so far from leading the

---

5   *Zoonomia*, vol. 1, 2nd edition, London, J. Johnson, 1796, p. 1.
6   Ibid. p. 537.
7   Ibid.

mind to atheism, would strengthen the demonstration of the existence of a Deity, as the first cause of all things'.[8]

Erasmus Darwin's view of nature as a series of causes and effects also led him to an idea of evolution, albeit more poetic than rigorously scientific. It is expressed in his poem, *The Temple of Nature*, where we read:

> *Organic Life beneath the shoreless waves*
> *Was born and nurs'd in Ocean's pearly caves;*
> *First forms minute, unseen by spheric glass,*
> *Move in the mud, or pierce the watery mass;*
> *These, as successive generations bloom,*
> *New powers acquire, and larger limbs assume;*
> *Whence countless groups of vegetation spring,*
> *And breathing realms of fin, and feet, and wing.*
> (Canto I, lines 295–302)

Erasmus Darwin famously described Unitarianism, not unkindly, as 'a feather-bed to catch a falling Christian', which is usually taken to mean that he thought of himself as having moved beyond it. Interestingly, however, a hymn attributed to him and included in James Martineau's *Hymns for the Christian Church and Home* (at number 137) and *Hymns of Praise and Prayer* (number 110) shows that this poetic physician was not the hard-headed atheist that is sometimes suggested. It begins:

> *The Lord, how tender is his love!*
> *His justice how august!*

and continues in verses two and three:

> *He showers the manna from above,*
> *To feed the barren waste;*

---

8   Ibid.

Or points with death the fiery hail,
  And famine waits the blast.
He bids distress forget to groan,
  The sick from anguish cease;
In dungeons spreads his healing wing,
  And softly whispers peace.

Perhaps this explains why not all Unitarians have disowned even Erasmus Darwin, let alone his grandson. *The Christian Life*, in May 1913, prefaced its feature on 'Unitarian Scientists 1813–1913' with the following comment: 'There have been many famous scientific men of the Unitarian faith ... prior to 1813, such as Sir Isaac Newton, Henry Cavendish, John Dollond, Dr. Joseph Priestley, Erasmus Darwin, and others'.[9]

Charles Darwin wrote a biography of his grandfather, intending to set the record straight about a man whose memory was much abused and wilfully misrepresented because of his radical opinions. His religious beliefs in particular were (and are) misunderstood, as Charles was at pains to point out. Erasmus Darwin – like himself – had falsely been accused of atheism, along with the moral flaws that supposedly went with it. Charles's assessment of his paternal grandfather in this regard is summarised by Desmond and Moore as follows: 'His intellectual and moral qualities were outstanding, and he too was traduced as a radical atheist'.[10]

But although Erasmus was no atheist, and although he counted prominent Unitarians among his closest friends and associates, he was notoriously scornful about the organised religion of his time. Writing of a meeting with Darwin in 1796, Samuel Taylor Coleridge (1772–1834) relates:

---

9 *The Christian Life*, 10 May 1913, p. 253.
10 Desmond and Moore, *Darwin*, p. 636.

> *Dr. Darwin is an extraordinary man, and received me very courteously*
> *– He had heard that I was a Unitarian, and bantered incessantly on*
> *the subject of Religion ... When he talks on any other subject, he is a*
> *wonderfully entertaining and instructive old man.*[11]

Erasmus Darwin's complaint about the Christianity that he encountered was that it held people in intellectual subjection through 'credulity, superstitious hope and the fear of hell', the result being a state that he regarded as a 'disease'. Religion – particularly of the 'enthusiastic' variety favoured by the Methodists – resulted in people having 'intellectual cowardice instilled into ... their minds from infancy', with 'credulity ... made an indispensable virtue'.[12] I expect the Unitarians of the day would have agreed with that, but for Erasmus Darwin they were still too closely tied to Christian ideas of supernaturalism and an interventionist God.

Charles Darwin's mother was **Susannah Wedgwood (1765–1817)**, an avowed Unitarian and daughter of the great potter, **Josiah Wedgwood I (1730–95)**, founder of the famous Etruria Works in Burslem, near Stoke-on-Trent. Josiah was one of the most prominent Unitarian laymen of his day, and he and his family attended the Unitarian Meeting House in Newcastle-under-Lyme. He was a friend and associate of **Joseph Priestley (1733–1804)**: minister, scientist, and the leading Unitarian theologian of the late eighteenth and early nineteenth centuries.

Priestley, along with both of Charles Darwin's grandfathers, was a member of the Lunar Society, so named because they met when the moon was full. This was a purely practical arrangement – it made evening travel easier. 'The Lunaticks', as they called themselves, included a number of Unitarians notable for their liberal, reforming attitudes on the great issues of the day – in particular the French Revolution and the abolition of slavery. It was Josiah Wedgwood I who, in 1786, made the famous anti-slavery medallion, with its cameo of a kneeling African in chains

---

11 Letters, I. 178–9; quoted in King-Hele, *Erasmus Darwin*, p. 41.
12 King-Hele, *Erasmus Darwin*, p. 55.

and bearing the legend, 'Am I Not a Man and a Brother?' These words are probably derived from Paul's Epistle to Philemon, verse 16, with its appeal for reconciliation and a radically transformed relationship of loving brotherhood, 'in the flesh and in the Lord', between a slave-owner and a runaway slave. This was the apostle's time-bomb under slavery. Finally, by the late eighteenth century, its inherent moral force was ready to shake slavery's foundations and, in the nineteenth, to precipitate its fall.

According to Priestley's *Memoirs*, however, 'neither politics nor religion ever were the subjects of our conversations'.[13] Rather, 'philosophy', meaning science and its practical applications, 'engrossed us wholly'. One idea that the Lunar Society discussed was Erasmus Darwin's rudimentary theory of evolution, set out in *Zoonomia* and the posthumously published *Temple of Nature*. Priestley did not involve himself in this conversation, however, as it would have encroached on the subject of religion. We know, however, that Priestley later rejected the elder Darwin's theory because it was in conflict with his own understanding of the Creator.[14]

Given this radical, 'freethinking', Unitarian background, it may come as something of a surprise to learn that, on 17 November 1809, his parents took the infant Charles to St Chad's Anglican church in Shrewsbury to be baptised. This reflected a pattern – by no means uncommon at the time – by which many Unitarian families outwardly and occasionally 'conformed' to the Church of England. In the early nineteenth century, with Britain at war with Revolutionary France, and the loyalties of Dissenters already suspect, one reason for such outward conformity was to establish the respectability of increasingly wealthy Unitarian families. 'Upwardly mobile', they had an aspiring social position to maintain. They became wary of being thought unduly 'radical', or even disloyal, in matters of religion and politics, especially after the 'Church and King'

---

13 Joseph Priestley, *Memoirs of Dr. Joseph Priestley, to the Year 1795. Written by himself: With a Continuation to the Time of His Decease, by His Son, Joseph Priestley*, Northumberland, Pennsylvania, 1806, cited in John Ruskin Clark, *Joseph Priestley: A Comet in the System*.
14 Clark, op. cit., p. 34.

riots that drove 'Gunpowder Joe' Priestley out of Birmingham and eventually out of the country. Furthermore, Unitarians and other Dissenters (along with Roman Catholics) were debarred from public office and from the two English universities under the terms of the notorious seventeenth-century Test and Corporation Acts. One simple way round the provisions of the Acts, however, was to be nominally Anglican, and to give your children the same option by having them baptised and perhaps confirmed in the Church of England – a compromise disapproved of by the more conscientious on both sides.

One Unitarian who saw such compromise by second- and third-generation Dissenters as a step towards a loss of their distinct identity was the educationalist, poet, and pamphleteer, **Anna Laetitia Barbauld (1743–1825)**. She wrote of their decline into 'spiritless indifference', as the wealth built up by the 'sobriety, industry and abstinence from fashionable pleasures' of their parents and grandparents made them 'eager to enjoy their riches' and 'to mix with that world' from which their Dissenting values had formerly guarded them. Once this has been set in train, says Barbauld, it is but a short step to a Dissenting sect abandoning its religious identity altogether: '... it must in a short course of years melt away into the establishment, the womb and grave of all other modes of religion'.[15]

This tendency was reinforced by another powerful factor. Until 1836, the Church of England also had a virtual monopoly on the solemnisation of marriage – only the Quakers were exempted. This humiliating injustice encouraged – indeed, virtually necessitated – 'occasional conformity' (a phrase in common use at the time) among Dissenters. The resultant ambivalent attitude of some Unitarian families towards the Church of England, and vice versa, was to feature at several points in Charles Darwin's story.

---

15   'Thoughts on the Devotional Taste, On Sects, and on Establishments' (1775), in *Anna Laetitia Barbauld: Selected Poetry & Prose*, eds. William McCarthy & Elizabeth Kraft, pp. 224–6.

Susannah Wedgwood Darwin worshipped in the Unitarian Chapel in Shrewsbury, where Samuel Taylor Coleridge had once preached. The poet was then a great admirer of Joseph Priestley, and was seriously considering entering the Unitarian ministry. This never happened, however, due in part, at least, to the fact that Susannah's brothers, Josiah II and Thomas, became Coleridge's patrons, so giving him sufficient financial independence not to have to earn a living. It was to services in this chapel that Susannah took her children, Marianne, Caroline, Susan, Erasmus, Catherine, and Charles. His earliest experience of worship was thus Unitarian. The same is true of his education. His teenage sister, Caroline, had been his first teacher; but when Charles was eight years old, Susannah sent him to the day-school conducted by the local Unitarian minister, George A. Case. But after Susannah's tragic death in 1817, Charles moved on to the 'great school' in Shrewsbury as a boarder. His brother, Erasmus, was already a pupil there.

The headmaster was the Revd Dr Samuel Butler, and the religious ethos of the 'great school' was Anglican, with daily worship in the chapel. Charles was a pupil there from 1818 to 1825. He did not shine academically, deriving little benefit from the school's heavily Classical curriculum apart from some of the odes of Horace, 'which I admired greatly'. His overall verdict, as delivered in his *Autobiography*, written many years later in 1876, was bleak: 'The school as a means of education to me was simply a blank.' He made many friends, however, 'whom I loved dearly'.

During those years, according to his *Autobiography*, he also enjoyed reading Shakespeare and the poetry of Lord Byron (whose wife, Anne Isabella Milbanke, was a Unitarian), Sir Walter Scott, and James Thomson, whose work *The Seasons* was immensely popular at that time. Darwin was to derive a great deal of pleasure from poetry in the earlier part of his life, although he later lost all taste for it. He credited *Wonders of the World*, a favourite book in his schooldays, with first giving him 'a wish to travel in remote countries'.

2 Maer Hall: 'Uncle Jos' and the Wedgwoods

In his youth, Charles Darwin spent a great deal of time at Maer Hall in Staffordshire, the home of his mother's brother, **Josiah Wedgwood II (1769–1843)**. To Charles he was 'Uncle Jos', and he was one of the most important influences on the young Darwin. In his *Autobiography*, Charles writes this of a man who was like a second father to him: 'He was the very type of an upright man, with the clearest judgement. I do not believe that any power on earth could have made him swerve an inch from what he considered the right course' (p. 19).

Josiah Wedgwood II was another of those Unitarians who, in order to claim a full place in society, made compromises with the Established Church. His household seems to have exemplified the mixture of Anglican practice with the beliefs and liberal values of Unitarianism that was blurring the boundaries of the old 'Radical Dissent'. Desmond and Moore characterise it (with particular reference to Emma Wedgwood Darwin, Charles's first cousin and future wife) as 'Anglicanised Unitarianism'.[16] In the words of the wife of Josiah II, **Elizabeth ('Bessy') Allen Wedgwood (1764–1846)**, writing with undoubted irony, it is 'better to conform to the ceremonies' of the Church, 'for one can never be quite sure that in omitting them we are not liable to sin'.[17] Their eight children were baptised and confirmed in the Church of England, but most, including Emma, were Unitarian by conviction. She remained so for the rest of her life.

The main theological influence on the Wedgwood/Darwin family through three generations was **Dr Joseph Priestley**, an old friend of

---

16 *Darwin's Sacred Cause*, p. 136.
17 Desmond and Moore, *Darwin*, p. 19.

Josiah Wedgwood I.[18] He was Unitarianism's principal theologian in the late eighteenth century and well into the nineteenth. His aim was to cleanse Christianity of what he called its 'corruptions', among which were the doctrines of the Trinity, Original Sin, and Eternal Damnation. His Unitarianism looked towards universal happiness in this world and the next. And he was also, of course, a scientist – a chemist – whose theology incorporated a materialist view of the universe.

Bearing in mind what is often said about Emma Wedgwood Darwin's supposed fear that Charles's soul would be damned eternally, it is worth emphasising Priestley's view on the subject. He writes in his *Memoirs*: 'Since ... not reasons of justice or equity could lead men to expect more than an adequate punishment, proportioned to their crimes, there was far from being any reason to imagine that future punishments would be eternal.'

Such a disproportionate response would be incompatible with God's nature as 'a just and righteous governor'.[19] Indeed, Priestley taught the universal salvation of all souls. In the words of a historian of American Universalism, Charles A. Howe, in his account of Priestley's arrival in America: 'Priestley was, in fact, a Universalist as well as a Unitarian in his theology'.[20]

Although the Wedgwood family connection with the Meeting House in Newcastle-under-Lyme was maintained, in Maer they attended St. Peter's parish church, of which Josiah Wedgwood II had become patron. As patron, Josiah II had the right to appoint the vicar, and the man whom he installed was his nephew, the Revd John Allen Wedgwood. He was, like Charles Darwin, a grandson of Josiah Wedgwood I. His own father, John Wedgwood, was a founder member of the Royal Horticultural Society. As vicar of Maer, John Allen Wedgwood presided at various family rites

18 Ibid., pp. 8–9.
19 John Ruskin Clark, *Joseph Priestley: A Comet in the System*, p. 220.
20 *The Larger Faith: A Short History of American Universalism*, p. 10.

of passage and, apparently, was able to conduct them in a manner that would not offend Unitarian sensibilities. These rites were to include the wedding of Charles and Emma on 29 January 1839 and, many years later, the funeral of Charles's brother, Erasmus, in 1881. So even if John Allen Wedgwood's own beliefs were in accord with orthodox Anglican doctrine (and I do not know whether they were or not), he seems at the very least to have practised a very liberal tolerance which reflected and accepted his Unitarian family background. Perhaps he was a model for the sort of benign, theologically easy-going, country clergyman that Charles might have been himself if he had ever been ordained to serve in the Established Church.

Not that religion of any kind much concerned the young Charles in those carefree days. Given his family background and his schooling thus far, he was probably content simply to go along with the blend of Unitarian beliefs and liberal Anglican practice that prevailed at Maer. As he reflected in his *Autobiography* many years later, 'I do not think that the religious sentiment was ever strongly developed in me'. With his four male cousins – Josiah III (1795–1880), Harry (1799–1885), Frank (1800–88), and Hensleigh (1803–91) – he much preferred shooting on the Maer estate. And back at the Hall he could enjoy the company of their sisters, Charlotte (1797–1862), Elizabeth (1793–1880) and, especially, Fanny (1806–32) and Emma (1808–96) – the 'dovelies'. Emma's religious interests, however, were already expressed in teaching in the village Sunday school, along with Elizabeth.

At Maer, for all the fun and games, Charles was exposed to the seriousness of a family deeply and religiously committed to the values of liberal, even radical, reform and opposition to the evils of the time, most especially slavery.

3 Edinburgh, 1825–1827: John Edmonstone and William Rathbone Greg

The time came when Charles would have to enter a profession, as his father had no intention of subsidising a lifestyle of shooting and idleness. He was to study medicine, like his father and grandfather before him. His brother, **Erasmus Alvey Darwin (1804–81)** – known in the family as 'Ras' – was already studying medicine at Edinburgh University. There was a long tradition of Dissenters studying at the Scottish universities, where the exclusive religious rules governing admission to Oxford and Cambridge did not apply. However, medicine did not suit Charles, and he lasted only two years, or 'sessions'.

His time in Edinburgh was by no means wasted, however, and he had opportunities to further an interest in the natural world that was awakened in him as a child. He was also exposed to the thinking of people who did not see that world in religious terms, as well as to new scientific thinking and – in the case of the increasingly influential phrenology – to pseudo-scientific ideas. (Phrenology, with its links to racism and slavery, was to be a troublesome irritant for Darwin throughout his career.)

One of his most significant friendships formed in Edinburgh was with the taxidermist and former slave, **John Edmonstone**, who had been brought from Demerara (in modern Guyana) and freed by his 'owner'. Edmonstone was a skilled and highly intelligent man, who taught Darwin taxidermy (an important skill for naturalists at that time) and whose company Darwin greatly appreciated. He was living proof that the racist stereotypes of black people propagated by the slave-owners and their supporters were nothing more than self-serving and, perhaps, self-deceiving lies. Not that Darwin would have needed much convincing on this point. He came from a religious and family background steeped in

Abolitionism. Unitarian ministers like his grandfathers' friend, Joseph Priestley, had denounced slavery from their pulpits, and Unitarian poets had made their contribution to the cause. When, in 1791, an Abolitionist motion proposed by Wilberforce and seconded by the Unitarian MP, **William Smith (1756–1835)**, had failed in Parliament, **Anna Laetitia Barbauld** wrote her scathing, furious, 'Epistle to William Wilberforce'. She begins:

> *Cease, Wilberforce, to urge thy generous aim!*
> *Thy Country knows the sin, and stands the shame!*
> *The Preacher, Poet, Senator in vain*
> *Has rattled in her sight the Negro's chain...*

And later in the same poem:

> *Where season'd tools of Avarice prevail,*
> *A Nation's eloquence, combined, must fail:*
> *Each flimsy sophistry by turns they try;*
> *The plausive argument, the daring lye,*
> *The artful gloss, that moral sense confounds,*
> *Th' acknowledged thirst of gain that honour wounds:*
> *Bane of ingenuous minds, th' unfeeling sneer,*
> *Which, sudden, turns to stone the falling tear ...*[21]

The Unitarian Liverpool MP, **William Roscoe (1753–1831)**, whose Abolitionist stance was to cost him his seat in Parliament, had also made his protest in two long poems. In 'Mount Pleasant' (1777) he writes:

> *Shame to Mankind! But shame to BRITONS most,*
> *Who all the sweets of Liberty can boast;*
> *Yet, deaf to every human claim, deny*
> *That bliss to others, which themselves enjoy:*

---

21   *Anna Laetitia Barbauld: Selected Poetry & Prose*, McCarthy and Kraft, eds., pp. 122, 123.

Life's bitter draught with harsher bitter fill;
Blast every joy, and add to every ill;
The trembling limbs with galling iron bind,
Nor loose the heavier bondage of the mind.[22]

In 'The Wrongs of Africa' (1787), Roscoe describes a 'deep freighted' slave-ship, setting out, 'with human merchandize', on the deadly 'middle passage' to the Americas:

One universal yell, of dread despair,
And anguish inexpressible; for now
Hope's slender thread was broke; extinguished now
The spark of expectation, that had lurk'd
Beneath the ashes of their former joys ...
*   ... Female shrieks,*
At intervals, in dreadful concert heard,
To wild distraction manly sorrow turn'd;
And ineffectual, o'er their heedless limbs,
Was wav'd the wiry whip, that dropp'd with blood.[23]

Samuel Taylor Coleridge, who had close links with the Wedgwood family, also used his poetry in the Abolitionist cause. In his 'Greek Ode on the Slave Trade' (1792) he denounces the supporters of slavery in no uncertain terms:

O you who revel in the evils of Slavery, O you
who feed on the persecution of the wretched,
wanton children of Excess, snatching your
brother's blood, does not an inescapable Eye
behold? Does not Nemesis brandish fire-breathing
requital? Do you hear? Or do you not hear?[24]

22 *The Poetry of Slavery*, ed. Marcus Wood, p. 55.
23 Ibid., p. 64.
24 Ibid., p. 206.

In the Wedgwood household the young Charles Darwin would surely have read or heard poems such as these. And he must also have come across the impassioned and highly controversial testimony of **Thomas Cooper (1791–1880)**, a young Suffolk-born Unitarian minister who had spent four years ministering to the slaves – as best he could – on the Jamaican sugar estate of Robert Hibbert. Hibbert, himself a Unitarian, is proof of the sad fact that far from all Unitarians supported what Desmond and Moore have called 'Darwin's Sacred Cause'.

Cooper's *Facts Illustrative of the Condition of the Negro Slaves in Jamaica*, published in 1824, the year before Darwin went to Edinburgh, provided important evidence for the Abolitionist cause and aroused the ire of the Hibbert family and the whole slave-owning interest. In it Cooper catalogues the brutalities and base immorality of the slave system from first-hand experience and makes it clear that slavery and its operators bear the responsibility for the slaves' degradation, not the slaves themselves. Dealing with one common prejudice – that the slaves are inherently lazy and deceitful – Cooper writes: 'The unwillingness of the slaves to work, is proverbial; and how can any one expect them to be industrious? Idle or active, their wages are the same: they have no rational motive for exertion, but, on the contrary, every motive for deceiving the driver' (p. 41) – the slave-driver, that is, with his whip. Significantly, bearing in mind some later debates in which Darwin became involved, Cooper makes specific reference to the 'common opinion' among Jamaica's slave-owners 'that the Negroes are an inferior species' (p. 16). This was not an opinion which either Cooper or Darwin shared.

As far as religion was concerned, Desmond and Moore tell us that 'Heresy was rampant in Darwin's years in Edinburgh'.[25] They describe two of his most significant teachers there, Robert Knox and Robert Edmond Grant, as 'irreligious'. For Darwin, however, this atmosphere would not have been such a shock as it must have been for many. As

25 *Darwin's Sacred Cause*, p. 41.

Desmond and Moore observe, 'But Darwin took all this heretical talk in his stride. Perhaps, coming from a line of freethinking males and Unitarians, he considered it passé.'[26]

Among Darwin's friends and fellow-students at Edinburgh was the Unitarian, **William Rathbone Greg (1809–81)**. He shared Darwin's abhorrence of slavery, a conviction learned as a pupil of the Unitarian minister, Lant Carpenter. But Greg's view of black people, although benign, still consisted of conventional generalisations, unlike Thomas Cooper's more realistic account. As a noted writer on politics, economy, and theology, Greg was later to support the theory of 'evolution by means of natural selection', but without Darwin's optimism about human perfectibility. Despite their disagreement on this issue, they shared a similar view on the fate of indigenous peoples who found themselves in the path of 'civilisation'. Their only hope of survival, Greg later wrote, in *The Westminster Review* of 1843, was to become like the emancipated African 'cotton-pickers' of the West Indies, whom he described after a tour there as 'docile, gentle, humble, grateful, and commonly forgiving' and 'blessed with the virtues of peace, charity, and humility'. More ominously, Darwin was to write in *The Descent of Man*: '... the civilised races of man will almost certainly exterminate, and replace, the savage races throughout the world' (p. 183). In some respects, at least, Darwin's was the more realistic assessment, once we have decoded the terminology. However, this should certainly not be taken to mean that he approved of a tragic process that he had seen at first hand on his travels.

Greg belonged to a rich and influential Unitarian family who made their money in the cotton trade and were members at Manchester's influential Cross Street Chapel. His father, **Samuel Greg (1758–1834)**, built the famous Quarry Bank Mill in the Cheshire countryside, complete with a 'model village' for the workers, and a Unitarian chapel. The regime there was a combination of paternalism and authoritarianism. After

---

26 *Darwin*, p. 141.

University, William Rathbone Greg joined the family business and ran one of its cotton mills in the Lancashire town of Bury. A friend, John Morley, recalled that 'With his work people his relations were the most friendly, and he was active ... in trying to better their condition.'[27]

Greg's own account of his time in Bury gives a much less sunny picture. He wrote: 'I came here full of philanthropic visions and dreams, of brightening the intellects and purifying the character of the people committed to my care, but all these vanished before the anti-magical effects of a fortnight's residence among them'.[28] His sense of disillusion took its toll of his beliefs too, and Holt records that he 'lost his religious faith'[29] and became an agnostic.

27 Raymond Holt, *The Unitarian Contribution to Social Progress in England*, pp. 54–5.
28 Desmond and Moore, *Darwin's Sacred Cause*, p. 44.
29 Holt, op. cit., p. 189.

4 Cambridge, 1828–1831

His second son having failed at medicine, the freethinking Dr Robert Darwin decided that Charles should go to Cambridge University and then, like a very high proportion of its graduates, become a clergyman in the Church of England. In his *Autobiography*, Charles recalled his reaction to this proposal: '... from what little I had heard or thought on the subject I had scruples about declaring my belief in all the dogmas of the Church of England; though otherwise I liked the thought of being a country clergyman' (p. 20).

This suggests that his own beliefs still owed at least something to the Unitarian component of his family background, so why was the Unitarian ministry never considered? Perhaps because it would have required a higher degree of personal and religious commitment than Darwin was able or willing to give. In any case, very few Unitarian pulpits could guarantee a security of income and tenure comparable with that of a comfortable country living in the Church of England. The degraded condition of the minister of a Dissenting congregation was summed up by someone who knew from experience what she was talking about: Anna Laetitia Barbauld, the daughter, wife, and friend of Unitarian ministers.

Their minister, that respectable character which once inspired reverence and affectionate esteem, their teacher and their guide, is now dwindled into the mere leader of public devotions; or lower yet, a person hired to entertain them every week with an elegant discourse. In proportion as his importance decreases, his salary sits heavy on the people; and he feels himself depressed by that most cruel of all mortifications to a generous mind, the consciousness of being a burden upon those from whom he derives his scanty support.[30]

30 Anna Laetitia Barbauld, *Thoughts on the Devotional Taste, on Sects, and on Establishments*, revised edition, 1792, in W. McCarthy & E. Kraft, eds., *Anna Laetitia Barbauld: Selected Poetry & Prose*, 2002, pp. 225–6.

The life of the Anglican country parson, with plenty of free time to pursue his own interests, and its ready-made social standing, was an altogether more attractive prospect. Others in Charles's extended family had already taken that path.

So Cambridge it was to be. And in order to deal with those 'scruples', he read some books of Anglican theology, 'with great care'. The result was, as he wrote in his *Autobiography*: '... as I did not then in the least doubt the strict and literal truth of every word in the Bible, I soon persuaded myself that our Creed must be fully accepted' (p. 20) – a rather puzzling statement, it must be said, given his family background.

With something less than a sense of vocation, Darwin set aside his 'scruples', prepared to swallow the Church of England's Thirty-Nine Articles of Religion, and went off to Cambridge. While there, his most enjoyable studies had little to do with theology. He much preferred the lectures and field-trips of the Professor of Botany, John Stevens Henslow. Henslow, like most Cambridge academics an Anglican cleric, became a long-term friend of Darwin, although he could not accept his theory of evolution when it was finally made public. Another of Darwin's favourite pursuits at Cambridge was beetle-collecting in the nearby fens.

One theological book that did impress Darwin while at Cambridge was William Paley's *Natural Theology* (1802). This put forward an idea analogous to what is now called 'intelligent design', with every species individually created to play its own precise part in a perfect and harmonious Divine Creation. These species did not change, because, with their own pre-determined role in God's plan, they had no need to, even if they could have. In the sheltered clerical enclave of Cambridge, no one was likely to challenge this theologically acceptable view of nature, and Paley could give vent to effusions such as the following: 'In a spring noon, or a summer evening, on whichever side I turn my eyes, myriads of happy beings crowd upon my view'.[31]

---

31 Desmond and Moore, *Darwin*, p. 90.

Darwin was to blow this fanciful notion apart in *The Origin of Species*, but at the time, in Cambridge, he did not question Paley's premises. He was, as he later recalled, 'charmed and convinced by the long line of argumentation' (*Autobiography*, p. 22). Reflecting on his three years at Cambridge, Darwin wrote in the same memoir (p. 20): '... it seems ludicrous that I once intended to be a clergyman'. And that intention did not survive for long. As Darwin put it, it 'died a natural death ... when ... I joined the "Beagle"'.

Darwin was very happy at Cambridge. As far as religion was concerned, it was to have little real importance for him – other than learning doctrines that he was soon to abandon and even to vehemently reject. He was not untouched by the aesthetics of good Anglican worship, however, or, at least, its music. Of the anthem in King's College Chapel, he wrote: 'This gave me intense pleasure, so that my backbone would sometimes shiver.'

Unitarian connections were pretty much non-existent during these three years, of course, except on vacation visits home and to Maer, when other matters took precedence anyway.

5 The voyage of the 'Beagle', 1831–1836

HMS Beagle (centre) from an 1841 watercolour by Owen Stanley, painted during the third voyage while surveying Australia.

The invitation to join the round-the-world voyage of HMS Beagle as companion to Captain Robert Fitzroy (1805–65) was fortuitous, to say the least. In Darwin's final year at Cambridge he had read not only Sir John Herschel's *Preliminary Discourse on the Study of Natural Philosophy*, but also Humboldt's *Personal Narrative* – the German explorer's account of his travels in Central and South America. These, said Darwin in his *Autobiography*, 'stirred up in me a burning zeal to add even the most humble contribution to the noble structure of Natural Science' (p. 30).

The invitation came via Professor Henslow, who had suggested Darwin when he heard of the 'vacancy' on the Beagle expedition. Henslow knew well Darwin's passion for 'Natural Science', and also, perhaps, thought that his talents would be wasted in an obscure country parish. He urged

Darwin to take this opportunity. Charles's father was not keen at first, but the advice of Josiah Wedgwood II – 'Uncle Jos' – and the promise that the Admiralty would cover the costs soon persuaded him. The 'Beagle' set sail from Plymouth on 27 December 1831, only a few months after Darwin had graduated from Cambridge. As he was to recall in his *Autobiography*, 'The voyage of the Beagle has been by far the most important event in my life' (p. 36).

The voyage would also set in train the undermining of what were then his religious beliefs, but the young graduate who set out on the 'Beagle' had few qualms at the time. 'Whilst on board the "Beagle"', he recalled, 'I was quite orthodox' (*Autobiography*, p. 143). During the course of the voyage, when the opportunity presented itself, he was pleased to attend church services and to take communion. He was also very appreciative of the missionaries whom he encountered in various places, seeing them not only as a civilising influence on the 'natives', but also as their protectors against the worst effects of colonialism. He also had his long-standing family abhorrence of slavery painfully confirmed by witnessing it in operation in Brazil. It was an issue on which Darwin had serious differences with the High Church Tory captain of the 'Beagle'. In Argentina, his humanitarian instincts were further assaulted by seeing at first hand the effects of the genocidal war then being waged against the Native American population. And in the natural world too, Paley's harmonious vision was beginning to disintegrate, along with any idea that the Earth had been in existence for only a few thousand years.

Darwin the nascent scientist collected specimens of animals and plants both extant and extinct, and puzzled over the questions that they posed. He also collected rock samples and observed the geological formations of the places that he visited. He spent long periods travelling ashore, meeting an amazing variety of human beings and observing the natural settings in which they lived their lives. But few, if any, of the people whom he met were Unitarians, so as far as 'Unitarian connections' are concerned, there is little to say. The voyage was not without its spiritual interest, however.

His account of the voyage contains some passages which describe what might fairly be called religious or spiritual experiences. One notable example concerns the forests of South America:

> Among the scenes which are deeply impressed on my mind, none exceed in sublimity the primeval forests undefaced by the hand of man; whether those of Brazil, where the powers of life are predominant, or those of Tierra del Fuego, where Death and Decay prevail. Both are temples filled with the varied productions of the God of Nature: no one can stand in these solitudes unmoved, and not feel that there is more in man than the mere breath of his body.[32]

Here is Darwin writing almost as a 'nature mystic', reflecting the spirituality of the Romantic poets that he had read in his younger days – poets like the young Coleridge, who had once preached in Shrewsbury's Unitarian Chapel, and William Wordsworth. In reading Coleridge's 'Rhyme of the Ancient Mariner', some lines in particular might have spoken to Darwin, the seafaring naturalist:

> Beyond the shadow of the ship
>   I watch'd the water-snakes:
> They mov'd in tracks of shining white;
> And when they rear'd, the elfish light
>   Fell off in hoary flakes.
>
> Within the shadow of the ship
>   I watch'd their rich attire:
> Blue, glossy green, and velvet black
> They coil'd and swam; and every track
>   Was a flash of golden fire.
>
> O happy living things! no tongue
>   Their beauty might declare ...

---

32 *Voyage of the Beagle*, p. 374.

Wordsworth's *The Excursion* was to become Darwin's favourite poem for a time. Some lines from it must have seemed particularly appropriate as he reflected on his great journey:

> *So, westward, tow'rd the unviolated woods*
> *I bent my way; and roaming far and wide,*
> *Failed not to meet the merry Mocking-bird ...*[33]

It was the mocking-birds of the Galapagos that first gave Darwin the key to his great theory. Another passage seems to anticipate one that he was to write, about a very English experience, in the closing part of *The Origin of Species*. This one, however, concerns Australia:

> *I had been lying on a sunny bank, and was reflecting on the strange*
> *character of the animals of this country as compared with the rest of*
> *the world. An unbeliever in everything beyond his own reason might*
> *exclaim, 'Two distinct Creators must have been at work!' ... It cannot be*
> *thought so: one Hand has surely worked throughout the universe.*[34]

In his old age, however, Darwin would lament that such grand scenes would no longer 'cause such feelings to rise in my mind. It may truly be said that I am like a man who has become colour-blind' (*Autobiography*, p. 148).

During the voyage, Darwin had two books constantly at hand. Both, quite coincidentally, were by men with Unitarian beliefs. One was a work of science almost as seminal as *The Origin of Species* was to be thirty years later; the other was a work by one of England's greatest poets. This was *Paradise Lost* by John Milton (1608–74), and I wonder if his words, for Darwin, seemed to apply to the depredations already being wrought on the wonderful world round which he sailed? The fragility of the natural

---

33 *The Excursion*, Book 3, lines 944–6.
34 *Voyage of the Beagle*, p. 325.

world, or maybe of the contemporary beliefs about it, seems to apply to Milton's evocation of Eden before the Fall:

> ... About them frisking played
> All beasts of th' earth, since wild, and of all chase
> In wood or wilderness, forest or den;
> Sporting the lion ramped, and in his paw
> Dandled the kid; bears, tigers, ounces, pards
> Gambolled before them; th' unwieldly elephant
> To make them mirth used all his might, and wreathed
> His lithe proboscis; close by the serpent sly
> Insinuating, wove with Gordian twine
> His braided train, and of his fatal guile
> Gave proof unheeded ... [35]

The other book, *The Principles of Geology*, was by Sir Charles Lyell. Its three volumes first appeared as Darwin sailed round the world, and he managed to obtain copies as they were published. Lyell was to be Darwin's role model as a scientist, one of his greatest supporters, and, after Charles arrived home in October 1836, one of his closest friends.

---

35 *Paradise Lost*, Book IV, lines 340–9.

6 Sir Charles Lyell

The urbane and cultured baronet, **Charles Lyell (1797–1875)**, was, perhaps, the most important geologist of the nineteenth century. His *magnum opus*, the three-volume *Principles of Geology*, was first published in 1830–33 and went through many editions thereafter. On the later editions, Darwin's ideas and comments were to have a direct influence. By that time Lyell and Darwin were close friends and associates, sharing their thoughts on a wide range of subjects.

Lyell's primary contribution to geology was to establish the principle known as 'uniformitarianism'. This contradicted the idea that the Earth had been created in six days only a few thousand years ago (on 23 October 4004 BC, according to James Ussher, a seventeenth-century Anglican Archbishop of Armagh), as was still believed by many people. It also contradicted the scientific orthodoxy of the time, which perceived various major geological features as the results of one or more cataclysmic global events, such as the Old Testament Flood. Rather, Lyell argued, the multifarious geological features of the planet were the result of a variety of entirely natural – and mostly gradual – processes, operating over truly immense periods of time. This theory directly contradicted the prevailing literalist interpretation of the Genesis creation story. It fitted well, however, with Darwin's theory of evolution, which required immense periods of time for species to evolve. Darwin was to see the parallels between Lyell's challenge to the biblical account of creation and his own; as he wrote in *The Origin of Species*: 'I am well aware that this doctrine of natural selection ... is open to the same objections which were first urged against Sir Charles Lyell's noble views on "the modern changes of the earth, as illustrative of geology"' (p. 74).

During the period between Darwin's return from his voyage and the publication of *The Origin of Species* – some 33 years – he not only toiled ceaselessly on his theory of evolution, he also shared his ideas with a small circle of scientists, of whom Lyell was one, along with the botanist, Joseph Dalton Hooker (1817–1911); the biologist, Thomas Henry

Huxley (1825–95); and the physiologist, naturalist, and surgeon, **William Benjamin Carpenter (1813–85)**, who was a member of the Rosslyn Hill Unitarian Chapel in Hampstead, London.

Sir Charles Lyell; from Sarah K. Bolton: *Famous Men of Science* (New York, 1889)

Lyell was something of a mentor to Darwin, who was more than twenty years his junior. In his *Autobiography* Darwin writes of his old friend: 'I saw more of Lyell than of any other man, both before and after my marriage. His mind was characterised ... by clearness, caution, sound judgement and a good deal of originality' (p. 45).

Darwin took Lyell as his role model when undertaking his great study of 'variation in animals and plants, under domestication and nature'. He wrote in his *Autobiography*: 'it appeared to me that by following the example of Lyell in Geology ... some light might be thrown on the whole subject' (p. 56). Of Lyell as a scientist and writer, Darwin praises 'the wonderful superiority of Lyell's manner of treating geology, compared with that of any other author' (pp. 36–7). He adds a little later: 'he felt the keenest interest in the future progress of mankind. He was very kind-hearted ...' (p. 46).

During their long association Darwin and Lyell did not always agree, although even when their differences were considerable, their friendship does not seem to have been threatened. One of these differences concerned slavery, which Darwin had witnessed in Brazil, and Lyell had seen in America. Both men wished an end to slavery, not least because both firmly believed that all human beings were of one species – what Lyell called the 'great human family'. It seems odd, then, that this matter should have been a cause of dispute between them.

Slavery had been abolished in the British Empire while Darwin was on his travels, and by this time it was slavery in America that was at issue: the 'great crime', as Darwin called it.[36] The problem was this: whereas Darwin took an uncompromising, unequivocal Abolitionist stance, Lyell had been influenced by those whom he had met in America who argued for a 'moderate', gradualist approach. They claimed that the immediate emancipation of hundreds of thousands of uneducated, unskilled

36 *The Descent of Man*, p. 142.

slaves would bring chaos. And furthermore, some of the hospitable, cultured, and apparently humane people who welcomed and entertained Lyell and his wife, Mary, in the Southern States were themselves slave-owners. Thus Lyell was seduced into thinking that slavery could be left to wither away, without any help from 'extremist' Abolitionists in the North. Darwin was incensed, accusing Lyell of having more sympathy for those who owned the slaves than for the slaves themselves.

Nevertheless, despite their differences on this fraught subject, Lyell, who had heard **Theodore Parker (1810–60)** preach in Boston, gave Darwin a book by this militantly Abolitionist Unitarian minister. Parker's sermons cannot have been comfortable listening for Lyell, although they would have delighted Darwin. Parker was a strong supporter of the 'underground railroad': the network of sympathisers that conveyed escaped slaves to freedom in Canada, and he had former slaves in his congregation. He was prepared to defend by force anyone in his care who was threatened with capture under the terms of the notorious Fugitive Slave Law, originally enacted in 1793 but renewed and much strengthened in 1852. At a ministers' meeting in Boston, he declared:

> *I have in my church, black men, fugitive slaves. They are the crown of my apostleship, the seal of my ministry. It becomes me to look after their bodies in order to 'save their souls.' ... I have been obliged to take my own parishioners into my house to keep them out of the clutches of the kidnapper ... I have written sermons with a pistol on my desk, loaded ... and ready for action ... with a drawn sword within reach of my right hand.*[37]

In his writings and sermons Parker lambasted those (especially ministerial colleagues) who, by submitting to, or even supporting, the iniquitous provisions of the law, were '... willing to send their mother into slavery, pressing the Bible into the ranks of American sin'.[38]

37 Quoted in David B. Parke, ed., *The Epic of Unitarianism*, p.115.
38 Desmond and Moore, *Darwin's Sacred Cause*, p. 235.

As in Britain, Unitarian poets in America wrote passionately in support of the Abolitionist cause, among them **James Russell Lowell (1819–91)**, who was to feature in the last act of Darwin's life story. In his clarion call, 'The Present Crisis' (1843), one verse reads:

> *Slavery, the earth-born Cyclops, fellest of the giant brood,*
> *Sons of brutish Force and Darkness, who have drenched the earth with blood,*
> *Famished in his self-made desert, blinded by our purer day,*
> *Gropes in yet unblasted regions for his miserable prey; –*
> *Shall we guide his gory fingers where our helpless children play?*[39]

The other point of serious disagreement between Darwin and Lyell is also surprising, bearing in mind Lyell's unstinting support for Darwin's work, and his urging that the reluctant Darwin should publish his theory before someone else did. The point was that, initially, Lyell had grave doubts about 'evolution by natural selection'. He could not dispense with the need, as he saw it, for the hand of God at some point in the process – even if it was only to give humanity its spiritual component, the impetus to become truly human. His opposition weakened, however, as time went on, even if he never did quite 'go the whole Ourang' (as he quipped to Thomas Huxley) with Darwin about humanity's ape ancestry. Nevertheless, in a letter dated 18 May 1860, Darwin told Alfred Russell Wallace: 'Lyell keeps as firm as a tower, and this autumn will declare his conversion, which is now universally known'.[40]

It remained a cause of frustration to Darwin that Lyell – who had always counselled him to avoid controversy – continued to sit on the fence in public when, in private, he professed his 'conversion'. Hooker said of Lyell that he was 'half-hearted and whole headed',[41] but Lyell defended himself to Darwin in a letter dated 11 March 1863:

---

39 *The Poetry of Slavery*, ed. Wood, pp. 560–1.
40 *Evolution: Selected Letters of Charles Darwin*, Burkhardt, Evans, and Pearn, eds., p. 10.
41 Ibid., p. 84.

> *My feelings ... more than any thought about policy or expediency, prevent*
> *me from dogmatising as to the descent of man from brutes, which,*
> *though I am prepared to accept it, takes away much of the charm from*
> *my speculations on the past relating to such matters.*[42]

It had been a hard journey for the former Anglican. Lyell wrote to Darwin on 16 January 1865 about the personal impact of *The Origin of Species*: 'I had been forced to give up my old faith without thoroughly seeing my way to a new one'.[43] But in his *Autobiography*, Darwin says of Lyell's faith: 'He was ... thoroughly liberal in his religious beliefs, or rather disbeliefs; but he was a strong theist' (p. 45).

Lyell had not only become 'a convert to the Descent theory' (*Autobiography*, p. 46): he had also, at some point, become a Unitarian. As Raymond Holt tells us: 'The hearers of Martineau at Little Portland Street between 1859 and 1873 included Sir Charles Lyell (a regular member), Charles Darwin (a frequent visitor) ... and Professor W.B. Carpenter.'[44]

---

42 Ibid., p. 75.
43 Ibid., p. 119.
44 Raymond Holt, *The Unitarian Contribution to Social Progress in England*, p. 345.

7 Emma Wedgwood Darwin

Watercolour portrait of Emma Wedgwood Darwin, late 1830s, by George Richmond

Of all Charles Darwin's Unitarian connections, none was closer or more enduring than his relationship with **Emma Wedgwood Darwin (1808–96)**. She was his first cousin, the daughter of Josiah Wedgwood II, and, from 29 January 1839, his wife. She is often portrayed as a rather dim, simple-minded, 'evangelical' Christian; an unquestioning member of the Church of England who feared that Charles was destined for eternal damnation because of his ideas and his questioning of the literal truth of the Genesis Creation myths. This picture is, I think, very far from the truth, to put it mildly.

Emma was a life-long Unitarian who, for reasons related to her family's upwardly mobile social status, was baptised, confirmed, and, indeed, married in the Church of England. At the family home in Maer, Staffordshire, she lived in a dynamic intellectual atmosphere of radical, reforming politics and liberal Unitarian Christianity. She was intelligent and widely read, and well able to hold her own with Charles and his friends in London after their marriage, and at Down House, their marital home. The fact that she disagreed with Charles on matters of science should not be taken as meaning that she did not understand what he meant. And nor should her grave concerns about the religious implications of Charles's work be taken to mean that she seriously believed that he might spend eternity in hell.

Darwin's ideas were already well advanced when he and Emma were married, and yet a letter that she wrote to him a few days before the wedding hardly indicates that she regarded their religious disagreements, however serious, as having infernal implications:

> *There is only one subject in the world that ever gives me a moment's uneasiness and I believe I think about that very little when I am with you and I do hope that though our opinions may not agree upon all points of religion we may sympathize a good deal in our feelings on the subject.*[45]

45 Desmond and Moore, *Darwin*, p. 279.

What are more likely to have been her actual fears were expressed in another letter written at around the same time. First of all she puts forth a distinctly Unitarian view, and one that she held for the rest of her life: 'My reason tells me that honest and conscientious doubts cannot be a sin'.[46] 'Reason' and 'conscience' were Unitarian watchwords, and no Unitarian would have believed that eternal damnation (even if there were such a thing) awaited anyone because of their 'honest and conscientious doubts'.

So what *did* Emma fear? She continues: '...but I feel it would be a painful void between us'. This is not about eternal damnation or the fear of hell, but rather the threat posed to their future happiness as man and wife by disagreement on such an important matter as faith. She wrote of her fear 'that our opinions on the most important subject should differ widely'. In the same letter Emma famously urged Charles

> ... to read our Saviour's farewell discourse to his disciples which begins at the end of the 13th Chapter of John. It is so full of love to them and devotion of every beautiful feeling. It is the part of the New Testament I love best.

We do not know how far Emma intended Charles to read, but the full passage extends from chapter 13: 2 to chapter 16: 33, and its message is one of comfort and reassurance, inclusiveness – 'In my Father's house are many mansions' (14: 1) – and 'good cheer' (16: 33). It has been suggested,[47] on the grounds of 15: 6, that Emma also meant the passage to be a warning about hell. This is the verse about 'the branch that is withered', which is gathered up and 'cast into the fire and ... burned'. Personally, I would need more evidence to show that Emma, in recommending the whole passage, actually intended that this one atypical verse should outweigh all the rest. It is possible that she could have been influenced by Church of England dogma in this regard, but

46 Ibid., p. 270.
47 Desmond and Moore, *Darwin*, p. 270.

it should be borne in mind that the Anglican vicar of Maer, John Allen Wedgwood, was, to judge by his accommodation of Unitarian beliefs at Emma's wedding to Charles, himself very liberal in his theology. The fact of the matter is that Unitarians of all stripes did not believe in eternal damnation.

In the early part of the nineteenth century, the predominant Unitarian theology was that of **Joseph Priestley**, and, as Desmond and Moore state, he influenced 'three generations of intermarried Darwins and Wedgwoods'.[48] So what did Priestley say about eternal damnation? In his 'Forms of Prayer and Other Offices for the use of Unitarian Societies' (1783), Priestley states quite clearly: '... the Rational Dissenters do not think that the future state of any man will depend upon his opinions, but only on his disposition of mind, and his conduct in life' (pp. 3–4).

The foremost Unitarian theologian of the nineteenth century was **James Martineau**. He too was an influence on Emma Darwin. Even though, according to Desmond and Moore,[49] she took Charles along to Anglican services in London after their marriage, it was to Martineau that she looked as a guide. They state:

> Emma's Christianity, in which Jesus's revelation of a future life is not to be proved from Scripture but simply to be believed through the transforming power of the Gospels on the individual heart, was being defended at the time by the Unitarian theologian James Martineau.[50]

Martineau represented a paradigm shift in Unitarian theology. Priestley's Unitarianism had been founded on the Bible and reason. His was a rational faith, but one which still looked to Scripture as its supreme authority. Even so, Priestley did not accept everything in the Bible as literally true, including the plainly mythological opening chapters of

48 Ibid., p. 8.
49 Ibid., p. 281.
50 Ibid., n. 3, p. 702.

Genesis. As a scientist and a rationalist, he concluded that the world could not have been created in only six days, because

> *... so much of interposition, and deviation from regular laws, was ... improbable, considering the slowness with which the course of nature proceeds.*[51]

It was also Priestley's view that

> *... the creation, as it had no beginning, so neither has it any bounds; but that infinite space is replenished with worlds ...*[52]

Bearing in mind Priestley's influence on early nineteenth-century Unitarians in general and the Wedgwoods in particular, this might mean that Emma actually shared this view of the creation. She would certainly have been aware of it. Priestley did, however, accept the historical authenticity of most of the Bible's miraculous and supernatural content, particularly with regard to the accounts of the ministry (but not the birth) of Jesus.

Martineau, however, influenced as he was by the new biblical criticism developing in Germany, no longer saw the Bible in this light. For him, the individual conscience was the supreme authority in religion; the biblical accounts of the miraculous and the supernatural, when they meant anything significant at all, related to inward spiritual experience rather than to external events. The same went for what, in 'orthodox' Christian theology, was spoken of in terms of 'damnation'.

Martineau wrote in his *Endeavours after the Christian Life*: 'we ... carry our own circumference of darkness with us: for who can quit his own centre,

---

51 *The Theological & Miscellaneous Works of Joseph Priestley (1817–32)*, quoted in John Ruskin Clark, *Joseph Priestley: A Comet in the System*, pp. 78–9.
52 Ibid.

or escape the point of view – or of blindness – which belongs to his own identity?'[53] This is what Emma may have feared. Martineau continues:

> *He who is not with God already, can by no path of space find the least*
> *approach: ... in vain plant him here or there – on this side of death or*
> *that: he is in outer darkness still; having that inner blindness which*
> *would leave him in pitchy night ...*

Martineau was talking about a spiritual – even an existential – state arising from the wilful misdirection of one's own soul to pride and self-ishness rather than to service. It was not about our honestly held beliefs and doubts, or the intellectual particularities of our theological positions. What Emma feared, I suspect, was that Charles's religious doubts would somehow compromise his essentially good and loving nature, not only creating that 'void' between them, but extending it into the afterlife. As it was, they did not do so. Of eternal damnation, however, Martineau had this to say in *The Rationale of Religious Enquiry*:

> *I am prepared to maintain that if ... the doctrines of the Trinity and the*
> *Atonement and everlasting torments were in the Bible, they would still be*
> *incredible; that the intrinsic evidence against a doctrine may be such, as*
> *to baffle the powers of external proof.* [54]

In short, Martineau discarded 'the doctrine of eternal torments', regarding it as 'an absurdity'. Emma would have known this too. Even after the move to Down House, she and Charles were frequent visitors to relatives in London, and they attended services at Little Portland Street Chapel, where Martineau was minister from 1858. In 1873, Emma's niece, Effie Wedgwood, the daughter of Hensleigh and Fanny Mackintosh Wedgwood, was married there (to Thomas 'Theta' Farrer). In his 'Short History' of the chapel (p. 25), one of Martineau's successors there, Henry

---

53 Second Series 1847, 1907, pp. 65–6.
54 *James Martineau: Selections*, ed. Alfred Hall, p. 62.

Shaw Perris, writing in 1900, tells us: 'Charles Darwin, a connection of the Wedgwood family, was a frequent visitor to the chapel they attended'.

Emma's strong belief in the afterlife was a particular comfort when her sister, Fanny, died. Whether or not she ever really feared separation from Charles in eternity, she certainly regretted that he was denying himself the comfort of such a faith. While conceding that eternal life 'cannot be proved', she implored him not to forsake what Jesus did, 'for your benefit as well as for that of all the world'.[55] Her own deep faith must have given her some comfort when she and Charles suffered the greatest tragedy of their lives: the death of their beloved ten-year-old daughter, Annie, in 1851 after a particularly distressing illness.

Annie died in Malvern, where she had been taken for treatment. Charles was at her bedside, and the experience of witnessing Annie's suffering is generally credited with destroying the last vestiges of his belief in God as a personal and loving being. Emma was at home in Downe, awaiting news, and Charles' sensitive letter to her betrays none of this. He begins:

I pray God Fanny's note may have prepared you. She went to her final sleep most tranquilly, most sweetly at 12 o'clock today. Our poor dear child had had a very short life but I trust happy and God only knows what miseries might have been in store for her.[56]

A little later he continues: 'I cannot remember ever seeing the dear child naughty, God bless her. We must be more and more to each other my dear wife.'

Whatever this letter may or may not say about Emma's and Charles's beliefs, it illustrates powerfully the loving bond that existed between them and which never failed, no matter what the test. We see this again in a letter that Emma wrote to Charles in June 1861, in which she returns

---

55 Desmond and Moore, *Darwin*, p. 281.
56 Edna Healey, *Emma Darwin: The Inspirational Wife of a Genius*, p. 208.

obliquely to the theme of Charles's fading beliefs. But it is not about his fate after death, but rather her sadness that he has denied himself the comfort and support of faith at a time of great suffering. She writes:

> I am sure you know I love you well enough to believe that I mind your sufferings as much as I should my own and I find the only relief to my own mind is to take it as from God's hand, and to try to believe that all suffering and illness is meant to help us to exalt our minds and to look forward with hope to a future state.[57]

Hoping to comfort him with what comforts her, she writes perceptively:

> When I see your patience, deep compassion for others ... and above all gratitude for the smallest thing done to help you I cannot help longing that these precious feelings should be offered to Heaven for the sake of your daily happiness. But I find it difficult enough in my own case ... It is feeling and not reasoning that drives one to prayer.[58]

After saying that she feels 'presumptuous' in writing thus, she goes on: 'I feel in my inmost heart your admirable qualities and feelings and all I would hope is that you might direct them upwards, as well as to one who values them above everything in the world'.[59]

Neither Emma's faith nor her love for Charles ever faded. Darwin's loss of what she recognised as religious belief still distressed her, however, as did some of the implications of his work. In the words of John Hedley Brooke, Andreas Idreos Professor Emeritus of Science & Religion at Oxford, in a lecture in 2008: 'Emma admitted late in life that some aspects of his writing had been painful to her – particularly the view that the moral sense had been the product of evolution'.[60]

---

57 *Evolution: Selected Letters of Charles Darwin*, ed. Burkhardt et al., p. 33.
58 Ibid.
59 Ibid., p. 34.
60 'Darwin and God: Then and Now', p. 3.

Nevertheless, by the time of her husband's death, she was reassured that none of this would exclude him from the love of God: 'A God of love ... would not cast out a good man who had so earnestly searched for truth'.[61] Edna Healey comments: 'The God she believed in would not have rejected a person that was so clearly good, but would have regarded him with the same compassion as she herself did'.[62] And that, of course, is very much the God of her Unitarian faith.

During the years at Down House, Emma regularly worshipped at the parish church and took her children there too. Edna Healey tells us, though, that she 'envied Fanny and Hensleigh the Unitarian church they attended in London'.[63] This was Little Portland Street Chapel. In church in Downe, as Desmond and Moore record: 'When the Creed came to be recited ... Emma kept faith with her heritage. The congregation turned to face the altar but she faced forward, refusing any truck with Trinitarianism.'[64]

The idea that she was an 'evangelical' hardly fits with her own comment on Evangelicals, who, she wrote, 'imagine they feel shame for an inherently sinful nature ... I remember the infants school at Kingscote shouting out so jollily "There is none that doeth good. **No not *one***".'[65]

As a Unitarian she had little time for the doctrine of Original Sin or inherited guilt – not to mention the 'eternal damnation' that supposedly goes with it. In Downe she clashed with the local High Church vicar, who tried to impose lessons about the Thirty-Nine Articles in the supposedly non-denominational village school. 'Anglicanised Unitarian' she may have been in some respects, but she remained very much a Unitarian nonetheless. After her death, her Unitarian allegiance was effectively recognised in her obituary in the local paper, which, while saying that

61 Edna Healey, *Emma Darwin*, p. 322.
62 Ibid., p. 342.
63 Ibid., p. 199.
64 *Darwin*, p. 403.
65 Edna Healey, *Emma Darwin*, pp. 333–4.

she was a 'liberal subscriber' to the village church, also stated that she was 'not a member of the Established Church'.[66]

At home, too, she kept faith with her Unitarian heritage. Her daughter, Henrietta, recalled that, notwithstanding Emma's parish-church attendance and taking of the sacrament, 'She read the Bible with us and taught us a simple Unitarian Creed, though we were baptized and confirmed in the Church of England'.[67]

Her body now rests in the churchyard in Downe, but not with Charles. Ironically, she shares the family grave with sceptical Erasmus, the brother whom she once regarded as a bad influence on her young husband.

One final point: bearing in mind the common misrepresentation of Emma Darwin, one should note that her reading matter reveals an informed interest in religious and philosophical matters. Among the many authors whom she read were the American Unitarian scholar, Andrews Norton, Dexter Professor of Sacred Literature at Harvard and author of *Statement of Reasons for Not Believing the Doctrines of the Trinitarians*; David Friederich Strauss, whose *Life of Jesus* was one of the most radical and controversial contributions to the nineteenth century's so-called 'quest for the historical Jesus'; and Ralph Waldo Emerson, American Unitarian philosopher and essayist. Of the latter, she wrote that she was 'wading through Emerson as I really wanted to know what transcendentalism means and I think it means that intuition is before reason (or facts). It certainly doesn't suit the Wedgwoods, who never have any intuitions.'[68]

66 Ibid., p. 341.
67 Ibid., p. 199.
68 Ibid., p. 332.

8 Harriet Martineau and the 'radical Unitarians'

There were Unitarians who were even more dismissive than James Martineau of concepts such as 'eternal punishment'. They have been dubbed the 'radical Unitarians', although this is not a term that was in use at the time. Emma Darwin disapproved of their more sceptical theological views, and she regretted their influence on Charles.

Among their number was James Martineau's sister, the writer and sociologist **Harriet Martineau (1802–76)**, with whom Charles's name was briefly linked before his marriage to Emma. Harriet gathered around her a circle that included his brother, Erasmus, to whom she became very close. Charles had much to do with these particular 'radical Unitarians', especially when resident in London between 1837 and 1842. At this time he was filling his notebooks with ideas that would eventually see the light of day in *The Origin of Species*. The group also included Emma's brother, **Hensleigh Wedgwood (1803–91)**, who, with his wife, **Fanny Mackintosh Wedgwood (1800–89)**, went to live next door to Charles and Emma, who had been married in 1839.

As Charles formulated his theories, he discussed them with this circle, including such important issues as whether New Testament morality actually had an evolutionary origin rather than a divine origin. Desmond and Moore tell us: 'By the autumn of 1838 he had thrashed out all the social and moral issues with Hensleigh, Harriet, and Erasmus, and worked them into his evolutionary formula'.[69]

Characteristically, the 'radical Unitarians' questioned or dismissed all the supernatural and miraculous trappings of religion, focusing solely

69 *Darwin*, p. 262.

on its ethical role in shaping personal behaviour and social reform. They were Charles's natural supporters, regarding scientific advance as part and parcel of the quest for a better society. Darwin's contention that the moral and ethical teachings of Christianity were the consequence of human evolution rather than divine inspiration was in tune with their thinking. Similarly, they agreed with the idea that the universe is a place of natural laws, with no room for divine interventions. As Desmond and Moore put it:

> [Harriet] Martineau's scientific attitude was typical of radical Unitarians. She saw nature as predictable, predetermined, invariant. It was subject to law and order, not the province of miracle.[70]

She saw the universe in materialist terms, regarding material as 'spiritually endowed' rather than existing separately from a distinct 'spirit world'.

More generally, 'radical Unitarians' were particularly concerned with education – especially for women – and they included a number of feminists and political activists. In London at this time they were often associated with the ministry of the Suffolk-born **William Johnson Fox** at South Place Chapel. This brand of radical, sceptical Unitarianism was very different from Emma's version of the tradition, which not only took its inspiration from Priestley and James Martineau but was also strongly influenced by Anglican practice. But although she sometimes worried about Charles being unduly influenced by the little group centred on Harriet Martineau, she could also laugh about them. As she wrote to Fanny Wedgwood: 'Poor Martineau seems going downhill with Hensleigh and Erasmus, so I hope you will stick by her'.[71]

70   Ibid., p. 217.
71   Ibid., p. 217.

9 William Benjamin Carpenter, James Martineau, and *The Origin of Species*

William Benjamin Carpenter (1885 or 1886); artist unknown

William Benjamin Carpenter (1813–85) came from a leading Unitarian family which also produced the distinguished ministers **Lant Carpenter** (his father), **Russell Lant Carpenter**, and **Joseph Estlin Carpenter**. His sister, **Mary Carpenter**, was an educationalist, social reformer, and campaigner for women's suffrage. William himself was a zoologist, botanist, physiologist, and surgeon. Like Darwin he had studied medicine at Edinburgh, and like Darwin he was a passionate opponent of slavery, attacking those who argued that there were several unequal human species, rather than one single species. Thus Carpenter and Darwin had much in common. Their friendship began in 1844, when Carpenter moved to London to become Fullerian Professor of Physiology at the Royal Institution. He also became Professor of Forensic Medicine at University College, London. He was one of those scientists who co-operated with Darwin during the long gestation of *The Origin of Species*. An active Unitarian, he was a member of Hampstead's Rosslyn Hill Chapel, and also an attender at Little Portland Street Chapel, whose ministers in this period included Edward Tagart and James Martineau.

When, on 26 November 1859, *The Origin of Species* was published, Carpenter was invited to write two reviews. The first, as we learn in a letter dated 10 April 1860 from Darwin to Charles Lyell, was for a professional journal. Darwin wrote: 'There is a very long review by Carpenter in Med. Chirurg. Review: very good and well-balanced but not brilliant'.[72]

The other review by Carpenter was for the Unitarian periodical *The National Review*. It appeared in the January–April issue in 1860,[73] and was something of a contrast to the reception given to *The Origin* in some other religious quarters. It is clear from the review that the attacks on Darwin had already begun. Carpenter writes contemptuously: 'To such as look upon this question from the purely scientific point of view, any theological objection, even to Mr. Darwin's rather startling conclusion, much more to his very modest premises, seems simply absurd.'

---

72 *Evolution: Selected Letters of Charles Darwin*, p. 7.
73 Vol. X, Article VIII, pp. 188 et seq.

In defending Darwin's scientific method, which drew parallels between the selective breeding of domestic animals and natural selection, Carpenter also enters the theological argument with relish:

> *Why, then, should Mr. Darwin be attacked for venturing to carry the same method of inquiry a step further; and be accused (in terms it takes no spirit of prophecy to anticipate) of superseding the functions of the Creator, of blotting out his Attributes from the page of Nature, and of reducing Him to the level of a mere Physical Agency.*

It might be thought that Carpenter was beginning to go off-message with this theological gloss on Darwin's work, but it should be remembered that, by his own testimony, Darwin was still a Theist at this time – and was never an Atheist, preferring the word 'Agnostic', newly coined by his friend T. H. Huxley, in later years. As he wrote himself in his *Autobiography* (p. 149):

> *Another source of conviction for the existence of God ... follows from the ... impossibility of conceiving this great and wonderful universe ... as the result of blind chance or necessity. When thus reflecting, I feel compelled to look to a First Cause having an intelligent mind in some degree analogous to that of man; and I deserve to be called a Theist. This conclusion was strong in my mind when I wrote 'The Origin of Species'.*

In a letter of 1879[74] Darwin states categorically:

> *I have never been an Atheist in the sense of denying the existence of a God.*

In his *Autobiography* (p. 149) we read what was probably his final position:

> *The mystery of the beginning of all things is insoluble by us, and I for one must be content to remain an Agnostic.*

---

74 Ibid., p. 139.

And when an Atheist tried to tell him that Agnosticism and Atheism were the same thing, Darwin was at pains to deny it (*Autobiography*, p. 154, n.1), on the grounds that Atheism is 'aggressive', whereas Agnosticism is not. It will be seen, therefore, that Carpenter's theological refutation of Darwin's critics in his 1860 review of *The Origin* was in no way inappropriate. Nevertheless, at one point in the *National Review* article, Carpenter does seem to come close to a harmonious view of nature somewhat reminiscent of William Paley's:

> ... we do not hesitate to say that the orderly and continuous working of any plan which could evolve such harmony and completeness of results as the world of Nature ... spreads out before us, is far more consistent with our idea of that Being 'who knows no variableness, neither shadow of turning' [James 1: 17] than the intermittent action of a power that requires a succession of interferences to carry out its original design in conformity with successive changes in the physical conditions of this globe.

But Carpenter is not concerned only with theology. He also recognises that, with Darwin, something of a sea-change has occurred in the scientific world:

> We are disposed to believe ... that Mr. Darwin and Mr. Wallace have assigned a vera causa for that diversification of original types which has brought into existence vast multitudes of species, sub-species, and varieties referable to the same generic forms; and we think that the weight of evidence is decidedly in favour of such an extension of this doctrine from the present to the past, as will enable us to account for the modification of specific types ... as we pass from one geological formation to another.

(Alfred Russell Wallace (1823–1913) was the naturalist who devised a theory of 'natural selection' independently, a development which had forced Darwin to stop dithering and publish.) Carpenter recognised the true significance of Darwin's insight:

*The history of every science shows that great epochs of progress are those
not so much of new discoveries of facts, as those of new ideas which have
served for the colligation of facts previously known into general principles,
and which have thenceforward given a new direction to inquiry. It is
from this point of view that we attach the highest value to Mr. Darwin's
work.*

The publication of *The Origin of Species* was a truly seminal event. It
changed our understanding of nature for ever. We know this with the
benefit of hindsight, but Carpenter, combining his own scientific knowledge with his Unitarian openness to truth, saw this at the time:

*The doctrine of progressive modification by Natural Selection
propounded by Mr. Darwin, will give a new direction to inquiry into
the real genetic relationship of species, existing and extinct; and it has a
claim to respectful consideration ...*

And Carpenter draws back from any suspicion that he might share
Paley's benign view of nature, while at the same time embracing the ultimate optimism of contemporary Unitarian belief and of Darwin himself:

*Nor is the least of its recommendations that it enables us to look at the
War of Nature ... as not only marked by suffering and death, but as
inevitably tending towards the progressive exaltation of the races engaged
in it ...*

Although it would certainly be wrong to suggest that all Unitarians
were immediately and entirely convinced by *The Origin of Species*, or
saw no more room for God than Darwin did, none responded with
the kind of hostility that emanated from some Christian quarters.
Most famous among Darwin's contemporary opponents was Samuel
Wilberforce, bishop of Oxford. Sadly, in some quarters the hostility still
exists. Darwin's bitterest enemies today are the disturbingly influential
religious 'fundamentalists', a surprisingly modern phenomenon. But
among Unitarians at the time, *The Origin* was recognised and accepted

as a valid contribution to science and to what we now call 'the free and responsible search for truth and meaning'.[75]

James Martineau (1805–1900) sounded a cautious note about Natural Selection, while still praising Darwin for his 'candour'. The great Unitarian theologian could not accept what he called 'the range given to chance' in 'Darwin's category of "accidental variation"'. He summarised Darwin's theory, as he understood it, in the following terms:

> The known species of organisms are the residue preserved in the
> competition for life by some casual advantage accruing to them by natural
> selection. This advantage is a prize turned up by the wheel of a vast
> lottery, with the peculiarity that its ticket was not made out and deposited
> there before, preordained to be drawn by someone; but formed and
> inscribed itself by the molecular experiments of the machine. No one can
> deny that the beneficial feature might thus arise ...[76]

For Martineau, however, there was a big 'but'! For him there just *had* to be some other factor at work. Nevertheless, he took Darwin seriously and, no doubt, had him in mind, with others, when he wrote:

> It cannot be denied that the architects of science have raised over us a nobler
> temple, and the hierophants of Nature introduced us to a sublimer worship.
> I do not say that they alone could ever find for us, if we knew it not, who it is
> that fills that temple, and what is the inner meaning of its sacred things; for
> it is not ... through any physical aspect of things ... but through the human
> experiences of the conscience and affections that the living God comes to
> apprehension and communion with us. But when once he has been found of
> us ... it is of no small moment that in our mental picture of the universe, an
> abode should be prepared worthy of a Presence so dear and so august.[77]

75 *Unitarian Universalist Association Principles & Purposes.*
76 *A Study of Religion*, vol.1, 1888; *James Martineau: Selections*, ed. Alfred Hall, pp. 44–5.
77 *Essays, Reviews & Addresses*, Vol. III, v; *The Seat of Authority in Religion*, vol. I (1890); *James Martineau: Selections*, ed. Alfred Hall, p. 47.

We live in an orderly universe, Martineau said, in which 'the constancy of creation is the direct expression of the good faith of God'.[78] The contribution of science is to enable us to live reasonably and reverently in that universe – but it does not have all the answers. Carpenter, and Darwin too, would have agreed with that.

When Darwin died, Carpenter paid rich tribute to his old friend at a meeting of the British and Foreign Unitarian Association; he proposed a motion, which was passed unanimously, 'applauding Darwin for "unravelling the immutable laws of Divine Government" and for shedding light on "the progress of humanity"'.[79]

A copy was sent to an appreciative Emma.

78 *A Study of Religion*, vol. I; *James Martineau: Selections*, ed. Alfred Hall, p. 44.
79 Desmond and Moore, *Darwin*, pp. 675–6.

10 Francis William Newman, Francis Ellingwood Abbot, and the Free Religious Association

In a letter dated September 1871, Charles Darwin wrote to 'Dr. F. E. Abbot, of Cambridge, U.S.' that, for health reasons, he did not feel 'equal to deep reflection, on the deepest subject which can fill a man's mind' (*Autobiography*, p. 140). Abbot had requested an article for the American periodical, *The Index*, which he edited. It was the organ of a body called the Free Religious Association. But health problems were not the only – or even the real – reason for Darwin's decision not to accede to the request. He also told Abbot: 'I feel in some degree unwilling to express myself publicly on religious subjects, as I do not feel that I have thought deeply enough to justify any publicity' (ibid.).

This was not to be the end of the matter, however. Of all the many requests for Darwin to disclose his opinions on religion, this was the one to which he was most sympathetic. And in the end, a statement written by Darwin did appear in *The Index*, and it continued to do so for several years. The reason was that *The Index* represented a brand of religion with which Darwin could agree.

By that stage in his life, Darwin was totally disillusioned with what was generally regarded as 'religion'. In particular, he had become vehemently antagonistic to the 'orthodox' Christianity which he had learned at school and studied at Cambridge, with the (albeit rather lukewarm) intention of becoming an Anglican clergyman. And, perhaps like grandfather Erasmus Darwin, he regarded the Unitarian Christianity of his mother, his sisters, his uncle, and his cousins – including his wife Emma – as a less than convincing alternative: a mere 'feather-bed to catch a falling Christian'.

Although, as he writes in his *Autobiography*, 'I was very unwilling to give up my belief', he had, nevertheless, done so:

> ... *the more we know of the fixed laws of nature the more incredible do miracles become, ... the Gospels cannot be proved to have been written simultaneously with the events, ... they differ in many important details, far too important ... to be admitted as the usual inaccuracies of eye-witnesses; – by such reflections as these I gradually came to disbelieve in Christianity as a divine revelation.* (p. 144)

He continues: 'Thus disbelief crept over me at a very slow rate, but was at last complete'.

In a passage that Emma would not allow to be published, but which was restored later (1958) by Darwin's grand-daughter, Nora Barlow, the full level of his antipathy is made clear:

> *I can indeed hardly see how anyone ought to wish Christianity to be true; if so the plain language of the text seems to show that men who do not believe, and this would include my father, brother and almost all of my friends, will be everlastingly punished. And this is a damnable doctrine.*

It should not be thought, incidentally, that Emma objected to this passage because she disagreed with his basic point about 'everlasting punishment', in which, as a Unitarian, she did not believe. It was simply that, as she put it shortly after Darwin's death, 'It seems to me raw'. She added (significantly, if we are not to misunderstand her, as so many seem to do): 'Nothing can be said too severe upon the doctrine of everlasting punishment for disbelief – but very few now would call that Christianity'. And certainly no Unitarian would do so. But for Darwin, even the liberal Christianity that then dominated Unitarianism could no longer hold him.

However, liberal Christian Unitarianism was not the only kind on offer. He had previously encountered Unitarians who took a different line. In the 1850s, for example, he was reading books – such as *The Soul, Its Sorrows and Its Aspirations, Phases of Faith,* and *History of the Hebrew Monarchy* – by **Francis William Newman (1805–97)**. Newman, the brother of Cardinal John Henry Newman, was a friend of Frances 'Fanny' Mackintosh Wedgwood, the wife of Emma's brother, Hensleigh.

Newman's work was much admired by, among others, the novelists **Elizabeth Gaskell (1810–65)**, a Unitarian, and her Anglican friend, Charlotte Bronte (1816–55). Elizabeth Gaskell, whose novels Darwin 'read and re-read till they could be read no more' (*Autobiography* p. 98), was another member of the Wedgwood clan. Her grandmother, **Catherine Wedgwood (1726–1804)**, had married **William Willetts**, minister of the Wedgwood family's Unitarian chapel in Newcastle-under-Lyme. Elizabeth herself married **William Gaskell (1805–84)**, minister of the influential Cross Street Chapel, Manchester. When Elizabeth was writing her biography of Charlotte Bronte, Fanny Wedgwood provided crucial assistance by copying out Charlotte's letters. It is from this biography that we learn of the two novelists' shared interest in Newman. Gaskell recalled, 'After breakfast, we ... went out on the Lake [Windermere] and Miss Bronte agreed with me in liking Mr. Newman's *Soul...*'.[80]

In *The Soul, Its Sorrows and Its Aspirations*, first published in 1849, Newman sets out to consider what he calls 'The Natural History of the Soul': 'The Soul is to things spiritual, what the Conscience is to things moral; each is the seat of feeling, and thereby the organ of specific information to us, respecting its own subject' (p. 8). By calling it an 'organ', Newman gives the soul a natural, rather than a supernatural, context, and although he also calls it 'a higher organ', he discusses it in terms that might have struck a chord with Darwin as he was working out his ideas. Newman writes of the soul: '... its diseases are more hidden and more embarrassing, and in consequence its pathology will assume an apparently disproportionate part of a true theology' (p. 9).

---

80 Elizabeth Gaskell, *Life of Charlotte Bronte*, p. 310.

Newman was Professor of Latin at University College, London, and had previously been a tutor at Manchester New College, where he taught aspiring Unitarian ministers. He occupied the theological position in the Unitarian spectrum known as Theism. Other notable Unitarian Theists at the time were the pioneer anti-vivisectionist, **Frances Power Cobbe** – said to have been 'almost aggressively neutral towards Christianity' – and **Richard Acland Armstrong**, who was preaching a 'Wordsworthian' faith at High Pavement Chapel, Nottingham.[81]

Newman rejected 'the dreadful doctrine of the Eternal Hell',[82] and saw the promise of eternal life in 'a full sympathy of our spirit with God's Spirit'.[83] This was, essentially, James Martineau's position, and the one that Emma Darwin held too.[84] But Charles Darwin found Newman's ideas unsatisfactory. In particular, he saw Newman's belief in the intuitive source of the 'religious instinct' as still locating its origin in God. This conflicted with Darwin's own growing conviction that it had evolved with human society. For this and other reasons, Newman had only confirmed and increased Darwin's religious doubts.

Twenty years after reading Newman's books, Darwin received a pamphlet from the United States entitled 'Truths for the Times', written by **Francis Ellingwood Abbot (1836–1903)**, a radical Unitarian minister. It was in two parts. One was entitled 'Modern Principles', while the other consisted of what were called 'Fifty Affirmations of Free Religion'. This provided Darwin with something that suited him down to the ground, advocating as it did a religion that was humanistic, optimistic, post-Christian, and free of all the doctrines that he had rejected.

Abbot's pamphlet also echoed the evidence-based avowal of the common origin and essential unity of the human species that Darwin had made

---

81 Harry Lismer Short, in *The English Presbyterians*, ed. Bolam et al., p. 273.
82 Newman, *The Soul, Its Sorrows and Its Aspirations*, p. 141.
83 Ibid., p. 143.
84 Desmond and Moore, *Darwin*, pp. 376–7.

in *The Descent of Man*, first published in 1871. The idea that there were several unequal human species, with distinct origins, had been widely propagated – especially in America – and Darwin himself wrestled with the question of how, why, and to what extent the 'races' or 'sub-species' of humanity differed. In *The Descent of Man*, however, having considered possibilities that are disturbing to the modern reader, he at least comes down on the side of a single human species with one common origin. He writes (on page 678):

> ... *all the races agree in so many unimportant details of structure and in so many mental peculiarities, that these can be accounted for only by inheritance from a common progenitor; and a progenitor thus characterised would probably deserve to rank as man.*

It is, I think, clear that – with characteristic caution – Darwin was expressing his conviction that all human beings have a common ancestry, that all belong to the species *homo sapiens*, and that the theory of evolution will, when the dust has settled, bear this out. He writes, with particular reference to the debate between those arguing for a single human species and those arguing for several: '... that when the principle of evolution is generally accepted, as it surely will be before long, the dispute between the monogenists and the polygenists will die a silent and unobserved death'.[85]

He was to reaffirm his view in *The Expression of the Emotions in Man and Animals* the following year:

> *I have endeavoured to show ... that all the chief expressions exhibited by man are the same throughout the world. This ... affords a new argument in favour of the several races being descended from a single parent-stock, which must have been almost completely human in structure ... before the period in which the races diverged from each other.*[86]

85 *The Descent of Man*, p. 210.
86 *The Expression of the Emotions in Man and Animals*, p. 329.

Abbot had read Darwin's more explicit application of evolutionary theory to human origins with growing enthusiasm, and wrote to him:

> *If I rightly understand your great theory of the origin of species, it contains nothing <u>inconsistent</u> with the most deep and tender religious feeling. It certainly conflicts with the popular notion of God, but it seems to me to harmonize thoroughly with the enlightened ideas concerning him held by all highly cultured minds of today ... and I for one feel that you have done a vast service to true religion by your labours.*[87]

Abbot was a leading figure in the Free Religious Association and was the editor of *The Index*, one of its two journals. The Free Religious Association had been formed in 1867 and was the result of a major dispute between radicals and conservatives within the American Unitarian denomination. The conservatives, identifying strongly with the denomination's roots in New England Congregationalism, demanded that it remain strongly, avowedly, and exclusively Christian in its Unitarianism. According to Conrad Wright, the radicals, while not necessarily repudiating Christianity altogether – although a number 'felt they no longer could conscientiously accept the Christian name' – wanted something altogether broader and more responsive to new insights.[88] As Wright puts it, these radicals were 'free spirits ... who refused to acknowledge for Christianity any special rank among the religions of mankind'.[89]

At its foundation, the stated purposes of the Free Religious Association were 'to promote the interests of pure religion, to encourage the scientific study of theology, and to increase fellowship in the spirit'.[90] Its initial membership, about half of whom were – and mostly remained – Unitarian ministers, included such prominent figures as Ralph Waldo Emerson, Louisa May Alcott, Octavius Brooks Frothingham, Samuel

---

87 Letter of 1871, quoted in *Darwin and God: Then and Now,* by John Hedley Brooke, p. 4.
88 Wright: *A Stream of Light: A Short History of American Unitarianism,* pp.72–3.
89 Ibid., p. 71.
90 Earl Morse Wilbur: *A History of Unitarianism in Transylvania,* England & America, p. 474.

Johnson, Samuel Longfellow, John White Chadwick, William J. Potter, and Felix Adler. Emerson spoke at the inaugural meeting, and a substantial proportion of the Association had, like him, strong connections with Transcendentalism. Others, however, such as Abbot, disliked Transcendentalism's intuitional approach to religion and constituted instead a 'scientific school'.[91] The Association, although its membership was never large, had a significant influence on the way that American Unitarianism was to develop in the twentieth century. As one assessment puts it, the Free Religious Association

> ... welded the humanistic revolt against Christianity into a single movement based on individual freedom of belief, the scientific study of religion, and the conviction that a single, universal spirit underlay all historic faiths ... It stimulated the transformation of Unitarianism from a Christian into a flexible and pragmatic theism.[92]

In the 'Fifty Affirmations',[93] there is an evolutionary view of religion itself, with an outmoded Christianity seen as giving way to Free Religion, which is about human progress and the realisation of human potential. As the fiftieth Affirmation puts it:

> Christianity is the faith of the soul's childhood; Free Religion is the faith of the soul's manhood. In the gradual growth of mankind out of Christianity into Free Religion lies the only hope of the spiritual perfection of the race.

(By 'race', of course, Abbot was referring to the whole human race, the species *homo sapiens*.)

The Affirmations include a number of statements that reflect Darwin's views. On the origin of religion, for instance, the second affirmation states:

---

91 *A Stream of Light*, p. 72.
92 David B. Parke, *The Epic of Unitarianism*, p. 122.
93 Ibid., pp. 123–5.

The root of religion is universal human nature.

Numbers 36 and 37 state:

The great faith or moving power of Free Religion is faith in man as a progressive being.

The great ideal end of Free Religion is the perfection or complete development of man, the race serving the individual, the individual serving the race.

Even what might be called the more specifically spiritual dimension of the Affirmations was well suited to accommodate Darwin's open-minded Agnosticism. Numbers 39 and 40 state:

The great law of Free Religion is the still, small voice of the private soul.

The great peace of Free Religion is spiritual oneness with the infinite One.

We can be confident that the Association's 'Truths for the Times', with its 'Fifty Affirmations', closely reflected Darwin's own beliefs in the 1870s because, in a letter to Abbot dated 24 June 1871 and quoted in Abbot's article, 'The Coming Empire of Science', Charles wrote:

I have now read 'Truths for the Times', and I admire them from my inmost heart; and I believe that I agree to every word.[94]

And Darwin went further than this private endorsement. He gave Abbot permission to publish it, slightly amended, in *The Index*. Darwin wrote:

I have read again 'Truths for the Times', and abide by my words as strictly true. If you still think fit to publish them, you had better omit

94 *The Index, A Weekly Paper Devoted to Free Religion*, vol. 2, issue 51, 23 December 1871.

'I believe', and add 'almost' to 'every word', so that it will run – 'and I agree to almost every word.' The points on which I doubtfully differ are unimportant; but it is better to be accurate.

For Abbot, getting Darwin to agree to publication was a great coup – all the more so because, with Darwin's permission, this endorsement appeared in every issue of *The Index* from 1871 to 1880, by which time Abbot was no longer the editor, and Darwin requested its removal.

In Abbot's view, 'The supposed conflict between science and religion is superficial and unreal, when both are properly conceived.' He characterised 'Truths for the Times' as 'an honest effort on the part of modern religion to meet modern science as friend', and continued: 'The importance, then, of Mr. Darwin's deliberate approval of the "Truths for the Times" lies in the fact that he is a man who by his genius has done more to extend the bounds of science than any other man living.'

Darwin's enthusiasm for *The Index* and the Free Religion that it promoted can be seen in a letter that he wrote to Abbot on 6 November 1871: 'I fully ... subscribe to the proposition that it is the duty of every one to spread what he believes to be the truth; and I honour you for doing so, with so much devotion and zeal' (*Autobiography*, p. 141). A few years later, when *The Index* was having financial problems, Darwin and his eldest son William responded with a 'generous gift' as 'a token of deep sympathy "in your noble and determined struggle" for free religion'.[95]

In the radical American Unitarians of the Free Religious Association, in 'Truths for the Times' and its 'Fifty Affirmations', in *The Index*, and in Francis Ellingwood Abbot, Darwin found a religion after his own heart. He became, in effect, an 'overseas member' of the Free Religious Association.

---

95 Letter to Abbot, 20 December 1875, Desmond and Moore, *Darwin*, p. 591.

Incidentally, Abbot was not Darwin's first contact in the Free Religious Association. One of its founder members, Charles Eliot Norton, the editor of the *North American Review*, had visited Down House with his wife in 1868. Norton was a close friend of one of Darwin's scientific correspondents and supporters, the American botanist, Asa Gray, Fisher Professor of Natural History at Harvard, who was staying with the Darwins at the time. It is likely, therefore, that Darwin had heard of the Free Religious Association from Norton before receiving 'Truths for the Times' from Abbot three years later.

The radicals of the Free Religious Association might have been expected to be supporters of Darwin's scientific theories, but in fact his ideas were favourably received by American Unitarians right across the theological spectrum in the second half of the nineteenth century. According to Conrad Wright: 'there is no significant difference ... between the reception of Darwinian evolution by Radicals and by conservatives'.[96] As the nineteenth century drew towards its close, the humanistic optimism that Darwin shared with the Free Religious Association became the shared spirit of American Unitarians in general. As Conrad Wright puts it:

> *More important for Unitarians at the close of the century than their*
> *long-standing differences was the common acceptance of an all-pervasive*
> *evolutionary optimism, so that conservative and radical alike could join*
> *in accepting at least the fifth of James Freeman Clarke's Five Points of*
> *Unitarian Belief: 'The progress of mankind onward and upward for ever'.*[97]

96 *A Stream of Light*, p. 93.
97 Ibid., p. 94.

11 Darwin: the Unitarian verdict

Charles Darwin, 1881 (the year before his death);
photograph taken by the studio of Elliott & Fry

Charles Darwin died at home, at Down House, on 19 April 1882. He had wanted to be buried in the village churchyard, where his brother, Erasmus, already lay. A simple rough-hewn coffin had already been ordered from the local carpenter. But this would never do, as far as his (by now) many influential friends and admirers were concerned. He was accorded what amounted to a virtual state funeral in Westminster Abbey, where his old friend, Sir Charles Lyell, had already been laid to rest.

The funeral took place on 26 April. Everybody who was anybody was there – with the notable exceptions of Queen Victoria; the Prime Minister, William Ewart Gladstone; and Emma Darwin – who mourned at home, away from the pomp and entirely inappropriate Anglican ceremony in the Abbey. Shortly beforehand she had written to her brother, Hensleigh Wedgwood, and his wife, Fanny: 'You will hear about Westminster Abbey, which I look upon as nearly settled. It gave us all a pang not to have him rest quietly by Eras; but William felt strongly, and on reflection I did also, that his gracious and grateful nature would have wished to accept the acknowledgement of what he had done.'[98]

Unitarians were well represented in the Abbey, not least by many of the 33 members of the Darwin and Wedgwood clan who were present, led by his eldest son William Erasmus Darwin. Another significant Unitarian presence in the Abbey came in the person of one of the pall-bearers, the US Ambassador, **James Russell Lowell**, fulfilling this role 'in grateful recognition of the interest taken by Americans in Mr. Darwin's works'.[99]

Although he was there to represent the American people in general, Lowell's presence was particularly fitting, in that he was a Unitarian. Perhaps the words of one of his own hymns, taken from his poem, 'The Present Crisis', seemed doubly appropriate for the occasion. Firstly, because it was originally written as an Abolitionist hymn: Lowell shared

98 Desmond and Moore, *Darwin*, pp. 669–70.
99 Ibid., p. 669.

Darwin's detestation of slavery. And secondly, because Darwin's theory of evolution by means of natural selection, so widely reviled in 1859, had, by 1882, been generally accepted, even by most of the churches. In the words of Lowell's hymn:

> Once to every man and nation
>   Comes the moment to decide,
> In the strife of Truth with Falsehood,
>   For the good or evil side ...
> Then to side with truth is noble
>   When we share her wretched crust,
> Ere her cause bring fame or profit,
>   And 'tis prosperous to be just;
> Then it is the brave man chooses,
>   While the coward stands aside,
> Till the multitude make virtue
>   Of the faith they had denied.

As far as evolution was concerned, the latter must have been true of quite a number of those present at Darwin's funeral in the Abbey! Incidentally, Lowell himself, a noted poet as well as a respected diplomat, has a small memorial in Westminster Abbey. He died in 1891.

The delicious irony of the formal proceedings was not lost on another American Unitarian, **John White Chadwick (1840–1904)** – hymn-writer, minister (in New York), and member of the Free Religious Association. With a piece of deliberately irreverent paraphrasing, he wrote triumphantly: 'The nation's grandest temple of religion opened its gates and lifted up its everlasting doors and bade the king of science come in'.[100]

In Britain, the Unitarian response was led by **William Benjamin Carpenter**, with the motion – unanimously adopted at a meeting of the

---

100 Ibid., p. 676.

British and Foreign Unitarian Association – praising Darwin for unravelling the 'immutable laws of the Divine government' and shedding light on 'the progress of humanity'.[101] At the meeting, as reported in the Unitarian newspaper *The Christian Life* on 29 April 1882 (pp. 202–3), Carpenter recalled how Darwin had been honoured for his contribution to science, and praised the characteristic humility with which he had held his theory of evolution, not hesitating to point out its problems (as yet unresolved) himself:

> *Dr. W. B. Carpenter truly observed at the meeting ... on Wednesday,*
> *that, quite independent of evolutionist views, fifteen years ago the Royal*
> *Society conferred its highest honour on Darwin for the immense ability,*
> *research, and discoveries which were clearly his own. Nor must we forget*
> *that Darwin himself, in that most wonderful book of his on the question*
> *at issue, admits that the greatest naturalists were not on his side, and*
> *sets forth himself difficulties to his theory we think insuperable.*

That editorial article, 'Darwin's Work and Influence', was particularly concerned with Darwin's impact on religious belief. Referring to the service in the Abbey, it declared:

> *Mankind believe not less sincerely in both natural and revealed religion*
> *now, than before the days of Newton; nor can we think of the large*
> *assemblage of both scientific and religious men, who met on Wednesday*
> *to pay their respect to the great mind that has passed on, without*
> *accepting that thousands and tens of thousands accept the theory*
> *Darwin has endeavoured to popularise in his book, 'The Origin of*
> *Species', without the least injury to their conviction of the truthfulness of*
> *the Christian religion.*

The article invokes Darwin himself to establish that his work need pose no problems for religious believers:

101 Ibid., pp. 675–6.

And we have it on record in Darwin's work that he was persuaded his views were not in conflict with religion. He says, 'To a reasonable mind the Divine attributes must appear, not diminished or reduced in any way by supposing a creation by law, but infinitely exalted.' ... He says again that his views of 'creation and the origin of species, appear to reflect far more wisdom and dignity on the Creator than the common views do.'

The article's author comments with reference to Darwin's beliefs (somewhat questionably, it must be said): 'We may or may not agree with these sentences, while from them it is transparent that the theory of Evolution in no way affected his conviction of the power, wisdom, and goodness of God.' But the message is clear. Darwin's science is fully reconcilable with religion: 'we deliberately affirm that neither faith, hope nor charity, are imperilled by Darwin's work. The facts of nature are but "Elder Scripture writ by God's own hand".' And the future is bright:

... we have in front of us a time when 'science will not vex religion nor religion vex science'. There can be no real hostility between religion with her heavenly purity, her queenly beauty, and science with its rounds of knowledge leading us to God.

Here was an expression of the optimistic spirit of nineteenth-century Unitarianism, with Darwin clearly regarded as one of its prophets. Another Unitarian newspaper, *The Inquirer*, took an equally enthusiastic line. Its editorial on 29 April 1882 declared: 'The 19th century ... has given to the world no grander name than that of Charles Darwin ... since Nicolas Copernicus, no man's works so strongly mark off an epoch of human thought ...'.

The Inquirer refers contemptuously to the initial opposition to Darwin's theory of evolution: 'The Titanic ravings which greeted the publication of "The Origin of Species" have now ceased, and public opinion ... is now unanimous in acknowledging the magnificence of the results which have sprung from the painstaking, patient observations of this one man.'

The article also contrasts the open-mindedness of those who were prepared to give Darwin a hearing with the attitude of the scientific and religious establishment back in 1859:

> ... it was curious to observe how much more spontaneously the new
> lights were followed even by those who had no special knowledge
> of biology, but whose minds were of the right type for appreciating
> the full value of the harmonious evolutionary system, than by those
> narrow specialists who could only think of difficulties involved in
> the interpretation of this particular weedy-flower, of that particular
> butterfly's wing.

The matter was not quite as settled, perhaps, as these confident tributes suggest. In a journal entry in 1894, **Beatrix Potter (1866–1943)** – herself a Unitarian – finds that Darwin is still a controversial subject in some households. On a visit to relatives in Stroud, when the conversation had turned in that direction, she writes of her cousin, Mary Hutton: 'Mary ... seems to be curious to discover whether I should be shocked with so much Huxley and Darwin'.[102] Potter was not shocked, of course, but she did wonder whether the scientific world-view might not prove 'a poor exchange' and an 'impossible creed' for the 'lower classes', who found great comfort in traditional beliefs, especially when faced with great suffering and death. Nevertheless, she concludes, 'Truth is truth'.

Although best known as the author and illustrator of children's books, Beatrix Potter took a keen scientific interest in the natural world, in which she was supported and encouraged by her uncle, the distinguished chemist, **Sir Henry Enfield Roscoe (1833–1915)**. A professor at Owen's College in Manchester, he was influential in its transition to

102 *The Journal of Beatrix Potter from 1881 to 1897*, new edition, 1989, p. 322.

University status. Roscoe belonged to a prominent Liverpool Unitarian family, and his grandfather, William Roscoe, had been an MP, a poet, and a leading Abolitionist. Beatrix Potter's family background was thus very similar to Darwin's.

Richard Acland Armstrong (1843–1905), the Unitarian minister and Theist, laid great stress on the spiritual value of poetry, especially that of William Wordsworth, although in his opinion the great poet had one rival in terms of his significance for nineteenth-century thought:

> [Wordsworth's] greatest utterances ... are the Bible of a new and larger faith which was to destroy the narrow creeds of established Christianity; they constitute ... the mightiest single intellectual influence of the nineteenth century; the only possible rival being that illuminating and penetrating conception associated with the splendid name of Darwin.[103]

And Armstrong pays further tribute to Darwin, placing him in the top rank of those born in the nineteenth century (which Wordsworth was not, of course): 'Alfred Tennyson was born in that year of wonder and of grace which gave birth to the three absolutely greatest Victorian Englishmen, Gladstone, Darwin, Tennyson'.[104]

Armstrong writes, in praise of Tennyson's poetry: 'let us add the wonderful lines on the Making of Man, in which Darwinian Evolution is made to minister to faith in the ultimate perfection of our race'.[105] The lines in question, which Armstrong reproduces, are these:

> Where is one that, born of woman, altogether can escape
> From the lower world within him, moods of tiger, or of ape?
> Man as yet is being made, and ere the crowning age of ages,
> Shall not aeon after aeon pass and touch him into shape?

103 Armstrong, *Faith and Doubt in the Century's Poets*, p. 26.
104 Ibid. pp. 69–70.
105 Ibid. p. 86.

All about him shadow still, but, while the races flower and fade,
Prophet eyes may catch a glory gaining on the shade,
Till the peoples all are one, and all their voices blend in choric
Hallelujah to the Maker: 'It is finish'd. Man is MADE'.[106]

On 10 May 1913, *The Christian Life* editorial (p. 305A), in stressing the importance of science for Unitarians, quoted words by John White Chadwick: 'Whatever else the effect of science on our theology, it has incalculably enhanced the force and value and significance of our doctrine of the Unity of God. There is no such Unitarian as science. There is no better Unitarian literature than ... Darwin's "Origin of Species".'

The same issue (p. 253) includes a feature on prominent Unitarian scientists of the previous hundred years. There, along with his friends, Sir Charles Lyell and William Benjamin Carpenter, and many others, is Darwin. His entry reads:

DARWIN, Charles Robert (1809–82), F.R.S., F.G.S., &c.; distinguished naturalist and author, grandson of Erasmus Darwin, poet and naturalist, and of Josiah Wedgwood, the famous potter – both of whom were Unitarians. Darwin's 'Life and Letters', published in 1887, affords some indication of what he owed to the influence of his early Unitarian training. His portrait by the Hon. John Collier adorns the walls of the Linnaean Society, and his statue by Boehm seems to preside over the Natural History Museum at South Kensington; but the impress of his work is stamped more indelibly than any other naturalist since the days of Aristotle, not only upon biology, but upon many sciences apparently least likely to be affected by it. Theology itself has been immensely affected by it.

At that time, Unitarians seem to have had no problem in regarding Darwin as 'one of their own', but later generations have been more hesitant. Raymond Holt, writing in 1938, in his influential work *The*

---

106 'The Making of Man', in *The Death of Oenone, Akbar's Dream, and Other Poems*, by Alfred Lord Tennyson (New York: Macmillan and Co., 1892), pp. 85–6.

Unitarian Contribution to Social Progress in England, fully acknowledges Darwin's Unitarian roots and their influence on his life, but does not regard him as a Unitarian. Since then there has been an unaccountable tendency on the part of Unitarians to put increasing distance between Darwin and themselves. This, I think, is unjustified. Unitarians – of various kinds – touched Darwin's life at so many key junctures from early childhood to old age that it seems perverse to deny the immense significance of their faith tradition for him. No other religious body can say as much. And no other religious body was better able to welcome his scientific insights and to accommodate his deepest beliefs, even as they changed. These days, everyone wants a piece of Darwin, often with very little justification. Unitarians may, out of deference to his own reluctance to say much on these matters, hesitate to 'claim' him themselves – but they should not be too quick to deny him, either.

Appendix

During Darwin's bicentenary year (2009), there was a debate among Unitarians about his significance for their faith tradition. Some responded to the occasion in worship. The Worship Panel of the General Assembly of Unitarian & Free Christian Churches published a pack of suitable materials, edited by David Dawson, which included the following, written by the Revd. Margaret Kirk. It is reproduced here with their kind permission. It also appeared in *The Inquirer* of 7 February 2009.

Litany – Thanks to Darwin
by Margaret Kirk

He looked with amazement and excitement at the world:
collected beetles from the fens of Cambridge;
barnacles, slugs, and polyps on the shore;
the fossilized bones of an ancient mammal, on his 'Beagle' voyage.
We are thankful that he did.

He watched phosphorescent creatures, corals and algae.
He wondered that so much beauty should be created.
We are thankful that he did.

This speculation took him beyond natural theology.
He would separate science from religious dogma.
He would study nature with a mind teeming with questions.
We are thankful he asked questions.

How did the damaged starfish with only three arms repair itself?
In God's beneficent world, why would a wasp sting caterpillars
and stuff them into its nest as food for its larvae?
We are thankful he asked questions.

How did a horse-sized mammal come to be extinct?
What could cause a whole species to die?
We are thankful he asked questions.

It took him another 20 years to find the confidence to publish his book,
The Origin of Species by Means of Natural Selection.
"What a fellow that Darwin is for asking questions."
We are thankful that he was.

Sources

Books and letters by Charles Darwin

Voyage of the Beagle (1839), edited and abridged with an introduction by Janet Browne and Michael Neve (Penguin Books, 1989).

The Origin of Species (1859), with an introduction by Jeff Wallace (Wordsworth Editions, 1998).

The Descent of Man, and Selection in Relation to Sex (2nd edition, 1879), with an introduction by James Moore and Adrian Desmond (Penguin Books, 2004).

The Expression of the Emotions in Man and Animals (2nd edition, 1890), edited by Joe Cain and Sharon Messenger, with an introduction by Joe Cain (Penguin Books, 2009).

Autobiography of Charles Darwin: with two appendices, comprising a chapter of reminiscences and a statement of Charles Darwin's religious views, by his son, Sir Francis Darwin (Watts & Co. 1929) (Icon Books, 2003).

Evolution: Selected Letters of Charles Darwin, 1860–1870, edited by Frederick Burkhardt, Samantha Evans, and Alison M. Pearn; foreword by Sir David Attenborough (Cambridge University Press, 2008).

Charles Darwin: The Beagle Letters, edited by Frederick Burkhardt, Sydney Smith, Charlotte Bowman, Janet Browne, Anne Schlabach Burkhardt, David Kohn, William Montgomery, Stephen V. Pocock, Anne Secord, and Nora Carroll Stevenson; with an introduction by Janet Browne (Cambridge University Press, 2008).

Darwin biography and commentary

Brooke, John Hedley, Andreas Idreos Professor Emeritus of Science and Religion at Oxford University, 'Darwin and God: Then and Now'; lecture given at St. Chad's Church, Shrewsbury, 27 February 2008.

Desmond, Adrian and James Moore, *Darwin* (Michael Joseph, 1991; Penguin Books, 1992).

Desmond, Adrian and James Moore, *Darwin's Sacred Cause: Race, Slavery and the Quest for Human Origins* (Penguin Books, 2009).

Healey, Edna, *Emma Darwin: The Inspirational Wife of a Genius* (Headline, 2001).

Padel, Ruth, *Darwin: A Life in Poems* (Chatto & Windus, 2009).

Erasmus Darwin

Darwin, Erasmus, *Zoonomia*, vol. 1, 2nd edition (J. Johnson, 1796).

Darwin, Erasmus, *The Temple of Nature* (J. Johnson, 1803).

King-Hele, Desmond, *Erasmus Darwin* (Macmillan, 1963).

Uglow, Jenny, *The Lunar Men: The Friends who made the Future 1730–1810* (Faber & Faber, 2002).

Contemporary Unitarian responses to Darwin

Abbot, Francis Ellingwood, 'The Coming Empire of Science', in *The Index, A Weekly Paper Devoted to Free Religion*, vol. 2, issue 51, 23 December 1871.

Carpenter, William Benjamin, review of *The Origin of Species* in *The National Review*, January – April 1860, Vol. X, Article VIII, pp. 188 et seq.

The Christian Life, 'Darwin's Work and Influence' (pp. 202–3), 29 April 1882.

The Christian Life, editorial (p. 305A), and 'Unitarian Scientists 1813–1913' (p. 253), 10 May 1913, Special Number, Vol. XXXIX, 1926–2707.

The Inquirer, editorial, 29 April 1882 (Obituary of Charles Darwin).

Potter, Beatrix, *The Journal of Beatrix Potter 1881–1897*, transcribed by Leslie Linder (1966), new edition with a foreword by Judy Taylor (Frederick Warne, 1989).

Slavery

Brady, Terence and Evan Jones, 'The Fight Against Slavery', British Broadcasting Corporation, 1975.

Cooper, Thomas, *Facts Illustrative of the Condition of the Negro Slaves in Jamaica: with Notes and an Appendix* (J. Hatchard and Son, 1824).

Priestley, Joseph, *A Sermon on the Subject of the Slave Trade; delivered to a Society of Protestant Dissenters, at the New Meeting in Birmingham; and published at their request* (J. Johnson, 1788).

Wood, Marcus, editor, *The Poetry of Slavery: An Anglo-American Anthology 1764–1865* (Oxford University Press, 2003).

Unitarian and Unitarian Universalist history

Bolam, C. G., Jeremy Goring, H. L. Short, and Roger Thomas, *The English Presbyterians: From Elizabethan Puritanism to Modern Unitarianism* (George Allen & Unwin, 1968).

Clark, John Ruskin, *Joseph Priestley: A Comet in the System* (Torch Publications, 1990).

Holt, Raymond, *The Unitarian Contribution to Social Progress in England* (George Allen & Unwin, 1938).

Howe, Charles A., *The Larger Faith: A Short History of American Universalism* (Skinner House Books, 1993).

Parke, David B., editor, *The Epic of Unitarianism: Original Writings from the History of Liberal Religion* (1957, Skinner House Books, 1985).

Perris, Henry Shaw, 'A Short History of Little Portland Street Chapel', 1900.

Wilbur, Earl Morse, *A History of Unitarianism in Transylvania, England and America* (Harvard University Press, 1952).

Wright, Conrad, *A Stream of Light: A Short History of American Unitarianism*, 2nd edition (Skinner House Books, 1989).

Unitarian thought, theology, and worship

Armstrong, Richard Acland, *Faith and Doubt in the Century's Poets* (James Clarke & Co., 1898).

Barbauld, Anna Laetitia, *Anna Laetitia Barbauld: Selected Poetry and Prose*, edited by William McCarthy and Elizabeth Kraft (Broadview Literary Texts, 2002).

Gaskell, Elizabeth, *The Life of Charlotte Bronte* (1857, J. M. Dent, 1908).

Martineau, James, *Endeavours after the Christian Life*, Second Series (1847, British & Foreign Unitarian Association, 1907).

Martineau, James, editor, *Hymns for the Christian Church and Home*, 7th edition (John Chapman, 1850).

Martineau, James, editor, *Hymns of Praise and Prayer* (Longmans, Green, Reader & Dyer, 1877).

Martineau, James, *James Martineau: Selections*, edited by Alfred Hall (The Lindsey Press, 1950).

Newman, Francis William, *The Soul, Its Sorrows and Its Aspirations* (1849, 7th edition, George Manwaring, 1862).

Priestley, Joseph, 'Forms of Prayer and Other Offices for the use of Unitarian Societies', 1783.

Priestley, Joseph, *Institutes of Natural and Revealed Religion*, vol. 1, 3rd edition, 1794.

Priestley, Joseph, *Memoirs of Dr. Joseph Priestley, to the Year 1795. Written by himself: With a Continuation to the Time of His Decease, by His Son, Joseph Priestley* (J. Johnson, 1809).

The Hymnbook Resources Commission, *Singing the Living Tradition* (Beacon Press /Unitarian Universalist Association, 1993).

Lightning Source UK Ltd.
Milton Keynes UK
09 March 2011

168975UK00001B/33/P